Technical and Vocational Education and Training: Issues, Concerns and Prospects

Volume 27

More information about this series at http://www.springer.com/series/6969

Rupert Maclean · Shanti Jagannathan
Brajesh Panth

Education and Skills for Inclusive Growth, Green Jobs and the Greening of Economies in Asia

Case Study Summaries of India, Indonesia, Sri Lanka and Viet Nam

50 YEARS ADB

Springer Open

Rupert Maclean
College of the North Atlantic-Qatar
Doha
Qatar

Brajesh Panth
Asian Development Bank
Mandaluyong, Metro Manila
Philippines

Shanti Jagannathan
Asian Development Bank
Mandaluyong, Metro Manila
Philippines

ISSN 1871-3041 ISSN 2213-221X (electronic)
Technical and Vocational Education and Training: Issues, Concerns and Prospects
ISBN 978-981-10-6558-3 ISBN 978-981-10-6559-0 (eBook)
https://doi.org/10.1007/978-981-10-6559-0

Library of Congress Control Number: 2017952531

ADB recognizes "Ceylon" as Sri Lanka; "Hong Kong" as Hong Kong, China; "Korea" as the Republic of Korea; "Vietnam" as Viet Nam; and "Hanoi" as Ha Noi.

Printed on acid-free paper

This Springer imprint is published by Springer Nature
The registered company is Springer Nature Singapore Pte Ltd.
The registered company address is: 152 Beach Road, #21-01/04 Gateway East, Singapore 189721, Singapore

Foreword

The rapidly growing attention to growth models that are environmentally sustainable and reduce carbon footprint drives developed and developing countries to accelerate the transition to a cleaner, greener growth. Some levers can make this transition faster and more enduring—and education, skills, and talent development is one such area.

This book is the outcome of a regional research project that explored the links between research, policy, and practice to enhance quality and relevance of skills development systems, particularly in industries with potential for high employment, growth, and greening. The book reinforces the mutual benefit of greening and moving up the global value chain for more inclusive growth.

Education, skills development, and tertiary education need retooling to build foundational knowledge, workforce training, and expand research and innovation in green economy-related fields. In the past decade, schools have promoted education for sustainable development by integrating environmental concerns in school curricula. There is also progress in research at university level on environmental sustainability and climate change. However, while the importance of skills for greening economies is acknowledged by both public and private sector representatives, Technical and Vocational Education and Training institutions have been slow to incorporate skills that facilitate the greening of various occupations and support the development of new occupations.

The transition to greener economies requires a pool of professional (e.g., architects and engineers) as well as vocational (mechanics and technicians) workers. Environment and sustainability-related courses need to be integrated into the curricula for vocational training. The availability of adequate talent, particularly for innovation, research, and development for a greener economy will foster new and innovative solutions and more cost-efficient and -effective technical approaches, financial models, and implementation capacities to achieve a successful transition.

Countries need to consider building the green growth knowledge base such as setting up national education initiatives for green and climate-resilient development and possibly a national climate education clearing house of knowledge. An example of how this can be done effectively is the Australian Green Skills Agreement, which seeks to build the capacity of the vocational education and training sector to deliver the skills for sustainability and required in the workplace.

For a greening environment, there is great potential in strengthening the links of environment, energy, education, and employment. At the Asian Development Bank, we will advance our efforts in strengthening these links by optimizing the role of education and skills development, especially as large-scale projects in clean energy, sustainable transport, climate change, and education, continue to be scaled up. These will help reap the benefits from holistic approaches as Developing Member Countries advance toward their sustainable development goals.

Bambang Susantono
Vice-President, Knowledge Management and
Sustainable Development
Asian Development Bank
Mandaluyong, Metro Manila, Philippines

Acknowledgements

This book is a product of a policy research study *Education and Skills for Inclusive Growth and Green Jobs* commissioned by the ADB on the role of education and training in responding to the transition toward green and sustainable economies and jobs. This regional research and development technical assistance project, which ran from 2011 to 2015, was initiated by the Sustainable Development and Climate Change Department (SDCC) of ADB to reinforce links between research, policy, and practice to enhance quality and relevance of skills development systems in India, Indonesia, Sri Lanka, and Viet Nam. We are grateful for the support provided by Carmela Locsin (formerly director general, SDCC), Woochong Um (formerly deputy director general, SDCC) and Bart Edes (formerly director, SDCC). We are grateful to the Education University of Hong Kong which was commissioned by ADB to undertake this project.

The project was conceptualized and led by Shanti Jagannathan, senior education specialist at the ADB. In addition to enabling research and analysis on the prospects and challenges of building dynamic skills development systems for a "greening" Asia, the project established the ADB International Skills Development Forum series, which has become a flagship knowledge-sharing event in education and skills in ADB, by financing four of the six annual International Skills Development Forums held so far.

Rupert Maclean (who at the time was at the Education University of Hong Kong) was the overall team leader for the project. In addition to leading the research in collaboration with international and national consultants, he led the development of a series of publications that include the book *Skills Development for Inclusive and Sustainable Growth in Developing Asia-Pacific* and a special edition of the journal Prospects on *Skills for Inclusive and Sustainable Development—Perspectives from the Asia Pacific Region and Beyond.* Special thanks are owed to Rupert Maclean for enabling extensive knowledge partnerships and publications arising from this project.

Brajesh Panth, chief of education sector group at the ADB led the overall coordination of the joint publication process with Springer.

This book draws on a wide range of primary and secondary source materials, as listed in the bibliography at the end of the publication. The authors drew heavily on the outputs from the above-mentioned research project. Four country reports and a regional report were prepared by Rupert Maclean in close collaboration with Saurabh Johri (technical and vocational education and training [TVET] consultant, New Delhi) and Will Douglas (Education University of Hong Kong).

The international research team for the project comprised Rupert Maclean (team leader), Margarita Pavlova (TVET providers specialist), and Belinda Smith (private sector skills specialist). The following national consultants included:

India: Saurabh Johri (TVET specialist and national team leader), Shruti Dasgupta (market research and survey specialist), and Shashikala Sitaram (gender specialist).
Indonesia: Punang Amaripuja (TVET specialist and national team leader) and several individuals supporting survey work and research on gender issues.
Sri Lanka: Uchita de Zoysa, (TVET specialist and national coordinator), Prasanna Ratnaweera (market research and survey specialist), and Bodiyabuduge Shani Ishika Weerasena (gender specialist). M. Saman Dassanayake provided support as a team leader and TVET Specialist during the initial stage of the study.
Viet Nam: Nguyen Quang Viet (TVET specialist and national team leader), Nguyen Thi Hoang Nguyen (market research and survey specialist) and Nguyen Thi Bich Thuy (gender specialist).

Dorothy (Petite) Geronimo, education officer and Sheela Rances, associate operations analyst from ADB provided extensive research assistance, program management, and administrative support throughout the project. Miguel Roberto Borromeo, associate information and research coordinator (education and health) provided research assistance.

Frédéric Asseline, principal climate change specialist at the ADB, peer reviewed, and provided significant suggestions to the final manuscript.

David Sorrell and Xie Qing (Jenny) in the EdUHK, provided editorial assistance for the final reports. Mitzi Legge, Benita Judith, and Teresa Lingat from the Office of Applied Research and Innovation, College of the North Atlantic-Qatar, helped to prepare the manuscript for publication.

Assistance is also acknowledged from Joan Abbott-Chapman, University of Tasmania, who reviewed and advised on the two survey instruments and interview schedules developed for the study; John Fien, Swinburne University of Technology; and Iris Bergmann, RMIT University, who contributed helpful insights about the Australian experience concerning green skills and sustainable and inclusive growth, which guided the conceptual framework for this study.

Series Editor's Introduction

This book reports on a major research project funded by the Asian Development Bank in Manila, and undertaken by the Education University of Hong Kong. It examines a topic that is of great importance not only to countries and communities in the rapidly developing Asia and Pacific region, but also to countries worldwide as they seek to achieve balanced economic growth that benefits all sectors of the community, is sustainable, and minimizes the adverse impact of such development on the environment.

The growing prominence of Asian economies and corporations, together with globalization and technological innovation, are leading to long-term changes in trade, business, and labor markets. Asia's economies have achieved remarkable growth rates, particularly over the past two decades. If Asia continues to grow on its recent trajectory, it could, by 2050, account for 51% of world gross domestic product (compared with 27% in 2010), with a sixfold increase in per capita income [Asian Development Bank, 2011]). With these long-term changes in trade, business, and labor markets, there is a rebalancing of power and influence in the region, and between the region and other parts of the world.

Within Asia and the Pacific, policy frameworks are being prepared to ensure that these growing economic benefits are inclusive by improving the quality and outreach of skills development employability and sustainable livelihoods. Countries are also taking action to ensure that economic growth in Asian and Pacific countries is "green," which minimizes adverse impacts on the environment. The transition to green growth currently focuses on efficient use of energy; greater use of renewable energy and the associated investment in technology development; waste reduction leading to lower pollutant emissions; production processes that conserve; the recycling and reuse of natural resources; and, an understanding that environmental regulations, standards, and economic instruments are not a hindrance to production, but are essential for achieving inclusive, sustainable, and environmentally friendly growth.

The adoption of green growth policies in support of green jobs and the greening of existing occupations are likely to require new skills to respond to the needs of the labor market. In this study, green jobs are defined as

> *Jobs that reduce the environmental impact of enterprises and economic sectors, ultimately to levels that are sustainable, are called green jobs. This comprises work in agriculture, industry, services, and administration that contributes to preserving or restoring the quality of the environment while also meeting the requirements of decent work, involving adequate wages, safe conditions, workers' rights, social dialogue, and social protection.*

This definition is consistent with the one used by the International Labour Organization and relevant others (ILO/UNEP/IOE/ITUC, 2008).

The major Springer book series in which this book is published examines issues, concerns, and prospects regarding education for the changing world of work. It seeks to provide research; evidence-based information about a diverse range of key, cutting edge aspects of technical and vocational education and training (TVET); and applied learning. The series showcases promising innovative approaches to TVET and education for the world of work. In doing so it also seeks to create an effective bridge among research, policy, and practice. This is a long-standing publications program that began in 2005 at the instigation of the UNESCO-UNEVOC International Centre for TVET in Bonn, Germany. The numerous volumes published to date in this major Springer book series provide a comprehensive, in-depth picture of current issues, concerns, and prospects in TVET, as they relate to both individual countries and worldwide.

July 2017

Rupert Maclean
Series Editor
College of the North Atlantic-Qatar

Contents

About the Authors

Rupert Maclean AO (Officer of the Order of Australia) is currently Qatar Petrochemical Company professional chair in Vocational Studies, and concurrently UNESCO chair on Technical and Vocational Education and Training and Sustainable Development, College of the North Atlantic-Qatar. He was previously chair professor of International Education, UNESCO chair in TVET and Lifelong Learning, and director of the Centre for International Education and Lifelong Learning at the Education University of Hong Kong. He is also the foundation director of the UNESCO-UNEVOC International Centre for TVET, Bonn, Germany. rupert.maclean@cna-qatar.edu.qa

Shanti Jagannathan is currently senior education specialist in the South Asia Department of the Asian Development Bank (ADB), Manila. She works on education policy and strategy for improving the quality, cost efficiency, and labor market relevance of skills development in ADB's developing member countries. Before joining ADB, she worked with the European Union as a development advisor on education reform programs in South Asia and on gender issues, democracy and human rights, and poverty alleviation. sjagannathan@adb.org

Brajesh Panth is currently Chief of Education Sector Group at ADB. He provides technical leadership to the Education Sector Group in ADB, leads the preparation of the Education Sector Group work plan, and facilitates collaboration across the sector and technical groups in ADB. He worked with the World Bank as senior education specialist before joining ADB in 2001. bpanth@adb.org

Abbreviations

ACB	ASEAN Centre for Biodiversity
ADB	Asian Development Bank
AICTE	All India Council for Technical Education
APEC	Asia-Pacific Economic Cooperation
ASSOCHAM	Associated Chambers of Commerce and Industry of India
ATI	Advanced Training Institute
ATMI (Indonesian acronym)	Academy for Manufacturing Engineering
BAPPENAS (Indonesian acronym)	National Board on Development Planning
BAU	Business-As-Usual
BDI (Indonesian acronym)	Industrial Training Centre
BLPT (Indonesian acronym)	Balai Latihan Pendidikan Teknik (Technical Education and Training Centre)
BPL	Below Poverty Line
BTL	BAC Thang Long
CEDEFOP	European Centre for the Development of Vocational Training
CEO	Chief Executive Officer
CGTTI	Ceylon-German Technical Training Institute
CII	Confederation of Indian Industries
CIPET	Central Institute of Plastics Engineering and Technology
CNG	Compressed Natural Gas
CO_2	Carbon Dioxide
CSP	Concentrated Solar Power
CSR	Corporate Social Responsibility
DMC	Developing Member Country

DTET	Department of Technical Education and Training
EC	European Commission
ECOTECH	Economic and Technical Cooperation
EdUHK	The Education University of Hong Kong
EGILAT	Experts Groups on Illegal Logging and Associated Trade
EMCO	Energy Management Company
ENG	Energy Working Group
EU	European Union
EWEC	East-West Economic Corridor
FAO	Food and Agriculture Organization of the United Nations
FDI	Foreign Direct Investment
FICCI	Federation of Indian Chambers of Commerce and Industry
GCF	Green Climate Fund
GDP	Gross Domestic Product
GDVT	General Department of Vocational Training
GHG	Greenhouse Gas
GINSI	Importers Association of Indonesia
GPEI	Indonesia Exporters Association
GTZ	Gesellschaft für Internationale Zusammenarbeit, (German Federal Enterprise for International Cooperation)
HDI	Human Development Index
HE-VTPs	Higher Education Vocational Training Providers
HM Government	Her Majesty's Government (UK)
HR	Human Resources
HRD	Human Resource Development
HRDWG	Human Resource Development Working Group
IDR	Indonesian Rupiah
IL&FS	Infrastructure Leasing and Financial Services Ltd.
ILO	International Labour Organization
IMF	International Monetary Fund
Indecon	Indonesia Ecotourism Network
IOE	International Organisation of Employers
ISO	International Organization for Standardization
IT	Information technology
ITI (Indian term)	Industrial Training Institute
ITUC	International Trade Union Confederation
KADIN (Indonesian acronym)	Indonesia Chamber of Commerce
KKNI (Indonesian acronym)	Kerangka Kualifikasi Nasional Indonesia (National Qualifications Framework)
LCREE	Low Carbon and Resource-Efficient Economy

LMIS	Labour Market Information System
MHRD	Ministry of Human Resource Development
MNC	Multinational Corporation
MOE	Ministry of Environment
MoEC	Ministry of National Education and Culture
MOET	Ministry of Education and Training
MOF	Ministry of Finance
MOHE	Ministry of Higher Education
MOIC	Ministry of Industry and Commerce
MOIT	Ministry of Industry and Trade
MOLISA	Ministry of Labour, Invalids, and Social Affairs
MoMT	Ministry of Manpower and Transmigration
MOSDE	Ministry of Skill Development and Entrepreneurship
MOSPI	Ministry of Statistics and Programme Implementation
MOT	Ministry of Transport
MOU	Memorandum of Understanding
MPI	Ministry of Planning and Investment
MSMEs	Micro, Small, and Medium Enterprises
NAITA	National Apprenticeship and Industrial Training Authority
NAMA	Nationally Appropriate Mitigation Actions
NCCC	National Council on Climate Change
NCPC	National Cleaner Production Centre
NEPC	National Environmental Protection Council
NGO	Nongovernment Organization
NO_2	Nitrogen Dioxide
NOS	National Occupational Standards
NOSSDC	NOSS Development Committee
NQF	National Qualification Framework
NSDA	National Skill Development Agency
NSDC	National Skill Development Corporation
NSQF	National Skills Qualification Framework
NSSD	National Strategy for Sustainable Development
NVQ	National Vocational Qualification
OECD	Organization for Economic Co-operation and Development
OHS	Occupational and Health Safety
OSD	Occupational Skills Department
PG	Postgraduate
PLC	Public Limited Company or Programmable Logic Control
PPP	Purchasing Power Parity or Public–Private Partnership
PROPER	Program for Pollution Control, Evaluation, and Rating
PSU	Public Sector Undertaking
PV	Photovoltaics

SAARC	South Asian Association for Regional Cooperation
SCE	Steering Committee on Economic and Technical Cooperation
SD	Sustainable Development
SEZ	Special Economic Zone
SHG	Self-Help Group
SHRDC	SAARC Human Resources Development Centre
SLSEA	Sri Lanka Sustainable Energy Authority
SMEs	Small and Medium-sized Enterprises
SMK (Indonesian acronym)	Sekolah Menengah Kejuruan (vocational secondary school)
SOM	Senior Officials Meeting
SSC	Sector Skill Council
STAR	Standard Training and Assessment Reward
TPTWG	Transportation Working Group
TVEC	Tertiary and Vocational Education Commission
TVET	Technical and Vocational Education and Training
UGC	University Grants Commission
UK	United Kingdom
UN	United Nations
UNDP	United Nations Development Programme
UNEP	United Nations Environment Programme
UNESCO	United Nations Educational, Scientific and Cultural Organization
UNIDO	United Nations Industrial Development Organization
US	United States
USA	United States of America
USAID	United States Agency for International Development
VBCSD	Viet Nam Business Council for Sustainable Development
VEIA	Viet Nam Environmental Industry Association
VET	Vocational Education and Training
VSS	Vocational Secondary School
VT	Vocational Training
VTA	Vocational Training Authority
VTC	Vocational Training Centre
WWF	Worldwide Fund for Nature

Abstract

International interest in sustainable and inclusive growth has gained impetus because of the recognition that the "brown" economy model does not address global issues such as growing inequality and social marginalization, environmental degradation, and resource depletion.

Evidence has shown that the "green" economy is a potential source of employment, can halt further productivity loss, and can address climate change and environmental degradation.

Broadly, three kinds of skills sets for green jobs can be identified. The first is general sustainability literacy, predominantly in the form of soft skills, awareness, and action competence. The second set includes occupation-specific science, technology, engineering, and mathematics (STEM) skills, while the third set encompasses leadership and management skills aimed at green transition. Many of these skills exist and are transferable, but need to be further strengthened and mainstreamed, or complemented with additional green concepts and practices. Others, such as broader STEM skills and leadership skills are missing. All present particular opportunities for the training sector.

Each of the four countries studied—India, Indonesia, Sri Lanka, and Viet Nam—as part of this Asian Development Bank (ADB)–Education University of Hong Kong (EdUHK) research study, presents an overview of main research findings concerning education and skills for inclusive growth, green jobs, and the greening of the economy for each country. This includes examples of government and business sector responses to the issues and challenges being addressed, and examples of how technical and vocational education and training (TVET) systems and institutions are addressing both the revision of curricula in the context of green growth dynamics and patterns of training and skills development for meeting demands.

The research study reported on in this book is unique. The findings, conclusions, and recommendations are based on primary data that were specifically collected for the study. Similar studies conducted in Asia have relied largely on the secondary sources of data, and report on existing research and related literature. By comparison, this study—in addition to reporting on existing research and related literature—surveyed TVET providers and business enterprises and examined survey responses of policy makers and practitioners on key aspects of education and skills for inclusive growth and the greening of economies. In addition, in-country workshops were held in each of the four countries to ascertain the views of key stakeholders in government, nongovernment organizations, members of the international development community, TVET providers, and members of the business sector. These workshops were organized to discuss key aspects of green growth, green jobs, inclusive growth, and the greening of economies, and to discuss the main findings emerging from the research.

In addition to reporting on research findings from India, Indonesia, Sri Lanka, and Viet Nam, this book examines cross-cutting issues, concerns, and prospects regarding education and skills for inclusive growth and green jobs for the four countries. These are critical themes and issues in the selected industry sectors triggering a demand for green jobs in the region; how industry is responding to those demands; areas impeding the transition from traditional to green practices; the importance of skills development, especially green skills, with regard to successful examples and the reasons for their success; the role of TVET in addressing industry needs; reasons for the slow response of TVET toward green skills, and key impediments; and, what works concerning initiatives from countries to fast-track reforms in TVET to facilitate inclusive growth and the greening of economies.

The research study provides recommendations to help manage the transition toward green and inclusive growth presents a suggested framework for implementation of potential strategies and policy initiatives, and examines areas for further research.

Summaries are provided in this book of the case studies undertaken for India, Indonesia, Sri Lanka, and Viet Nam.

The full, detailed case studies are available for viewing and downloading from www.cna-qatar.com/research/unesco-unevoc.

Chapter 1
Overview

Abstract The green economy is a new economic paradigm which seeks to achieve economic development, while at the same time protecting the environment and achieving sustainable economic and social development. This requires transitioning to green jobs and green skills, and to creating new jobs in relation to the greening of workforces. Green jobs are relevant across all key sectors: agriculture, manufacturing, building, transport, tourism, and renewable energy. Skills acquisition and enhancement have great positive implications for all aspects of education and training, and for businesses.

Keywords Green jobs · Green skills · Green economy · Green growth · Sustainability · Skills development priorities · Waste management and recycling · Renewable energy · Skills acquisition and enhancement · Skills toward sustainability · Energy transition · Smart cities · Environmental goods and services · Education and training

Green Economy and Green Growth

The green economy is a new economic paradigm that aims to achieve development while protecting the environment. It is considered as a way forward, purporting that there should be no conflict between economic progress and environmental protection. Greening the economy does not impede wealth creation, nor employment opportunities. Instead, relevant sectors can present opportunities for investment, growth, and jobs given the necessary conditions during transition. Transitioning to a green economy can help mitigate the adverse impact of rapid population growth in the depletion of natural resources (UNEP 2011). Furthermore, the latest OECD report *Investing in Climate, Investing in Growth* (2017), provides an analysis of how low-emission and climate-resilient development can be achieved without compromising economic growth, competitiveness, or well-being. The report provides a detailed discussion of just transition issues and how governments can manage such issues.

R. Maclean et al., *Education and Skills for Inclusive Growth, Green Jobs and the Greening of Economies in Asia*, Technical and Vocational Education and Training: Issues, Concerns and Prospects 27, https://doi.org/10.1007/978-981-10-6559-0_1

Although there is no one-size-fits-all model for green growth, and policies and strategies depend on the prevailing economic situation and reflect local contexts and preferences of countries, all countries—developed, developing, and poor—can adopt green growth and more inclusive development. Green growth can open up opportunities for enhanced productivity through more efficient use of natural resources and energy; for optimizing their value for innovation made possible by favorable policies that enable enhancing values of assets and protecting the environment; and for new markets brought about by emerging demands for green technologies, goods, and services, thereby creating new jobs (World Bank 2012, OECD 2011).[1]

Green Jobs and Green Skills

The expected transition toward a green economy and jobs is anticipated to result in displacement of workers and job losses, while new jobs will arise in relation to the greening of workplaces. This is a challenge that depends on how wide or narrow the gap is between the skills that current workers possess and the set of skills that a green economy and green jobs require.

Across various definitions, the primary objective is more efficient use of energy and natural resources, minimal waste and pollution, and protection of the environment and ecosystems. It is also equally important to point out that green jobs have to be decent work: good jobs with sufficient wages, safe working conditions, job security, reasonable career prospects, and worker rights (UNEP 2008). Otherwise, a hazardous job with an inadequate wage goes against the poverty-reduction principle that is dovetailed to green jobs and green growth.

In "Green Jobs: Towards Decent Work in a Sustainable, Low-Carbon World," UNEP (2008) listed four expected impacts of green growth on employment[2]:

- Additional jobs will be created (as in the manufacturing of pollution-control devices added to existing production equipment).
- Some employment will be substituted (as in shifting from fossil fuels to renewables, or from truck manufacturing to rail car manufacturing, or from landfilling and waste incineration to recycling).
- Certain jobs may be eliminated without direct replacement (as when packaging materials are discouraged or banned and their production is discontinued).
- Many existing jobs (especially such as plumbers, electricians, metal workers, and construction workers) will simply be transformed and redefined as day-to-day skill sets, work methods, and profiles are greened.

---

[1]For more extensive reading on the concept and origin of green growth and green economy, refer to *A Guidebook to the Green Economy* (2012) by the Division for Sustainable Development, United Nations Department of Economic and Social Affairs, https://sustainabledevelopment.un.org/content/documents/GE%20Guidebook.pdf.

[2]UNEP (2008). Green Jobs: Towards Decent Work in a Sustainable, Low-Carbon World, UNEP/ILO/IOE/ITUC, September 2008. Nairobi, Kenya. P3.

Filling positions in skilled green professions and trades will require development and provision of sufficient training programs in polytechnics and universities, among others. Workers will need support to adapt or transition in terms of skills; and management will need reorientation in terms of new perspectives, as well as new skills in order to optimize the new skills of their workers. Training and retraining programs should be based on a concrete framework and plan, which, in turn, should be based on sound studies and assessment within each economy, taking into consideration the level of economic development and state of transition into a green economy.

The development of modules and curricula should be a dynamic process informed by experiences and lessons learned from other countries. Each country should be able to establish a definition of green skills that can be operational in its context. For example, Australia defines green skills as skills for sustainability—i.e., "technical skills, knowledge, values and attitudes needed in the workforce to develop and support sustainable social, economic and environmental outcomes in business, industry and the community."[3] The Organisation for Economic Co-operation and Development (OECD) classifies green skills together with entrepreneurial skills under a third group of skills called converging skills (the first two skills groups being basic skills and advanced/knowledge-intensive skills); green skills are "specific skills required to adapt products, services or operations to meet adjustments, requirements or regulations designed to stem further climate change or adapt to the impact it is already having." (ILO 2011a) A broad range of knowledge and technical, managerial, and conceptual skills is indispensable during transition into a green economy. The United Kingdom has developed a green skills checklist, which is composed of ten broad groups of skills (tier 1) applicable across sectors. These are broken down into groups of general skills (tier 2) and more specific skills (tier 3) (ILO 2011a).[4]

1 Green Jobs Across Key Sectors

While pursuing green growth may cause job losses, new jobs will be created, especially skilled jobs in emerging innovations for green activities. In this case, facilitating the transition of workers to new, expanding sectors where enterprises use cleaner alternatives and provide environmental services should form an important part of plans and strategies toward a green economy.

Investing in green activities has huge job creation potential. The skills development response could be a two-pronged approach: (i) reorienting and/or retraining/

[3]Green Skills Agreement: An Agreement between the Australian Government and the State and Territory Governments http://www.ivet.com.au/cgi-bin/user.pl?download_file=1&file=17.

[4]ILO (2011a). Table 5.4: Green skills checklist (United Kingdom), pp. 104–105. http://www.ilo.org/wcmsp5/groups/public/@dgreports/@dcomm/@publ/documents/publication/wcms_159585.pdf.

upskilling existing workers to equip them with green skills applicable in enterprises across key economic sectors, such as infrastructure, agriculture, and manufacturing; and (ii) training and educating incoming workers to produce a fresh supply of skilled workers and professionals, especially for sectors directly involved in green activities like renewable energy.

Agriculture. The main green growth objective in agriculture is to boost productivity through resource-efficient technologies, agricultural inputs, and farming practices. The main challenge is reconciling traditional practices with green agricultural technologies, practices, and products. Partnerships with relevant non-government organizations, education institutions, demonstration farms, and small and medium-sized enterprises (SMEs) that provide green agricultural products are crucial.

Skills development priority: Ensuring a sufficient pool of professionals and specialists in the fields of agricultural engineering, research and development (R&D) (especially in irrigation, pest management, and soil fertility), and agro-ecology.

Manufacturing. The key areas in improving resource efficiency in manufacturing in a green economy are green innovations in designing and developing products, material and energy substitution, and modification and control of processes using new, cleaner technologies.

Skills development priorities

- training related to changes in production in heavy manufacturing industries, as well as in SMEs in the areas of production processes and management (leaning toward energy and resource efficiency, recycling, and waste management);
- upgrading of skills for technicians and workers to align with technological changes; and
- production of professionals and specialists in R&D and in revision of education curricula and training programs, eco-design, products and services, and assessment of environmental impact to promote cleaner processes and systems.

Building. Greening the building sector opens up opportunities for job creation and job transformation in the areas of new construction and retrofitting, as well as opportunities for increased production of green construction materials and products like appliances and equipment; expansion of renewable energy sources and generation; energy-efficient operation and maintenance (O&M); and other related activities such as recycling, waste management, and water and sanitation.

Skills development priorities

- training of workers and certified professionals in retrofitting buildings into green buildings and in landscaping, among others; and
- orientation of architecture and building engineering toward green designs and use of resource-efficient materials.

Transport. Shifting to more environment-friendly means of transport and improving vehicle and fuel technology to reduce negative effects, such as depletion of resources and pollution, are the priorities in greening the transport sector. Job opportunities will be in green transport infrastructure, green vehicles and their operations, and alternative fuels and other related technologies. For example, hybrid vehicles (plug-in electric and hydrogen/fuel cell-powered vehicles) that aim to reduce gasoline consumption, and efficient public or mass transport systems such as rail transit and energy-efficient bus systems are important in greening the transport sector, and they have great potential for green job generation (UNEP 2008).

Skills development priorities

- training in green transport infrastructure labor and O&M;
- production of more resource-efficient vehicles and their maintenance;
- R&D in alternative fuel technology;
- transport infrastructure design and engineering; and
- public transport network construction, O&M, and management.

Tourism. In a green economy, tourism is "sustainable tourism" that is maintained within the context of the local social, economic, and cultural environment. Being a human resource-intensive industry, tourism is among the world's largest job generators and often requires quick training and allows youth, women, and even migrant workers an easy entry into employment. Employment in related areas, such as energy, water, and waste services, also provides opportunities.

Skills development priorities

- training of local unskilled and semiskilled staff, and reorientation and training of cottage industries and small and medium-sized enterprises (SMEs) that provide related products and services;
- training and reorientation of managers and owners of related enterprises; and
- integration of green orientation in the curricula of tourism-related programs in technical and vocational education and training (TVET) and higher education.

Waste management and recycling. The greening of the waste management and recycling sector means a significant shift to the 3R approach (reduce, reuse, and recycle) and minimal use of waste disposal methods like incineration and landfills and waste to energy. This can be made possible by new technologies for collection, recycling waste, and producing energy from organic waste, among others.

Recycling promotes significant savings in resources. It generates employment and provides income to about 12 million people in Brazil, the People's Republic of China (PRC), and the United States (US). Developing countries have about 15 million people who depend on waste collection for their livelihood. However, waste management and recycling usually draw in workers with low educational attainment and having limited economic opportunities. And due to the often poor working conditions, most jobs in waste management and recycling cannot be called green jobs, because they hardly fall into the category of decent jobs. They often involve child labor and run contrary to the principles of occupational health and safety and social protection, among others (UNEP 2011).

Skills development priorities

- training in facilities management, including on-site waste management and disposal,
- training in recycling and disposal of waste
- training and reorientation of managers and workers on minimal production of waste and use of environment friendly packaging
- training to local community workers to increase knowledge of sustainable waste management techniques and building greater community awareness.

2 Jobs in Renewable Energy

Demand for, and consumption of, renewable energy has grown steadily in Asia and the Pacific, as well as worldwide. This trend is expected to be sustained as countries transition to green growth. Renewable energy is becoming cheaper, producing lesser air pollution, and emerging related innovations continue to help implement renewable energy projects (ESCAP 2015). Countries are adopting national sustainable energy policies and plans and setting concrete goals. The PRC, for example, targets producing 16% of its primary energy from renewable sources by 2020. Japan aims to create 1.4 million new jobs in environment-related fields under its Green Innovation Program (OECD 2011).

In 2010, employment in renewable energy recorded 3.5 million workers, mostly in Brazil, the PRC, Germany, Japan, and the US. Employment grew particularly in Asia and North America (UNEP 2011). Renewable energy generation and distribution created 6.5 million jobs worldwide in 2013, growing to 8.1 million in 2015. This may reach 24 million jobs by 2030. Another example from Indonesia and South Africa showed that by spending $1 million in energy sectors, clean energy investments generate about 100 jobs in Indonesia and 70 jobs in South Africa. (UNIDO and GGGI 2015). For these two countries, net employment gains for clean energy investments are substantial and can be a net positive source of job creation. In 2015, the top 10 countries in job creation included four in Asia (Bangladesh, the PRC, India, Japan), and Asia's share of total employment in renewable energy was about 60%, a 9-percentage point increase from 51% in 2013 (IRENA 2016).

Jobs in renewable energy are not confined to energy production per se. The value chain covers manufacturing and distribution of renewable energy equipment, renewable energy project development, construction and installation work associated with the development of renewable energy capacity, O&M of renewable energy facilities, and a range of cross-cutting activities that contribute to more than one of the other value chain stages. The largest number of workers in renewable

Table 1 Hard-to-fill occupations in renewable energy

Wind energy	Solar energy	Hydropower	Geothermal	Bioenergy
• Project developers • Service technicians • Data analysts • Engineers: electrical, computer, mechanical, construction	• Installers and maintainers of photovoltaic and solar thermal systems • Building inspectors	• Engineers: electrical and O&M • Technicians • Tradespersons • Sustainability specialists	• Engineers: geothermal • Trainers	• Engineers: R&D and design • Service technicians • Trainers

Source IRENA 2014

energy is in manufacturing of equipment, project development, construction, and installation. Employment is relatively smaller in O&M, which can last for the 20- to 30-year lifetime of the installation.

IRENA (2014) listed several occupations in renewable energy considered as "hard to fill" (Table 1). Most of them are high-skills jobs such as engineers and technicians.

Skill shortages in green jobs are often a result of the general inability of education and training systems to respond to the demands of the growing industry. This may be exacerbated by lack of universities and research centers offering relevant programs. Other barriers include students' choices of programs versus the skills in demand, the absence of incentives for employers and/or industry to invest in developing the transferable skills of their workforces, limited access of the disadvantaged to training in terms of time and finance, and probably the stereotyping of women against certain disciplines such as engineering (World Bank 2012).

3 Skills Development for Green Jobs

Skills shortages for green jobs could thwart government efforts toward green growth and targeted environmental outcomes. As in any other skills shortage situation, addressing an existing job-skills gap during a transition to a green economy requires government to partner with industries and employers. Companies need to shift to a more holistic perspective supportive of sustainable development. Aside from investing in skills upgrading, firms should also provide their managers the necessary reorientation and training to develop awareness and capacities to sufficiently utilize the staff/workers' newly acquired skills (UNEP 2011).

Overall, skills development strategies for renewable energy can be outlined as the following broad processes:

Aligning Skills Provision Strategies for Green Jobs with National Development Policies

Public and private sector investments in skills provision for green jobs should be aligned with and supported by sustainable development and other enabling policies. Skills development outlined in green policies should be in synergy with strategies to increase labor capital, worker education, and labor productivity (World Bank 2012). Special focus and investment in science, technology, engineering, and mathematics (STEM) must be highlighted to respond to the existing and anticipated demand in the renewable energy sector for R&D and engineering occupations, among others. At the outset, a thorough review of existing skills development policies should be done with the end in view of aligning higher education and TVET with green growth policies.

Assessing Skills Requirements for Green Growth

It is equally important to accord high priority and investments in determining and forecasting evolving skills needs to inform assessment of and any adjustment in occupational skills profiles and training programs. The results of skills needs assessment should feed into labor market information systems to support strategies to match skills supply and demand in renewable energy and green jobs in other key sectors. Also to encourage increase in the supply of trainers, social dialogue, and advocacy with education and training institutions, industries, and all other stakeholders in evaluating and planning for renewable energy skills development must be conducted.[5]

Skills Acquisition and Enhancement

As mentioned earlier, in transitioning to a green economy, skills development programs have to provide for (i) retraining, upskilling, or adaptation training of existing workers for enterprises across sectors such as infrastructure, agriculture, and manufacturing; and (ii) training of new workers and producing skilled workers and professionals, especially in sectors directly related to the renewable energy sector.

A lot of skills, even knowledge of sustainable materials, carbon footprinting skills, and environmental impact assessment, may not be inherently "green." They become green largely depending on the context wherein they are used. Skills in building construction become green when they are used in the creation of energy-efficient buildings. Skills in impact assessment become green when the results help create more resource-efficient knowledge, practices, or products. This

[5]At the international level, skills gap analyses are done by institutions like the International Labour Organization, United Nations Environment Programme, and The International Renewable Energy Agency. At the national level, skills gap analyses must be spearheaded by governments and partners to support enabling policies in the field of education and training. At the national level, government can do skills gap analysis through regular employment surveys or project-based assessments, accompanied with stocktaking of entities providing related training and education such as universities and TVET institutions.

means that traditional skills will remain relevant in the context of green growth, but will need to adapt to green practices, for example, new knowledge and practices in energy efficiency, when applied in green jobs and enterprises. While skills development strategies for green jobs will vary depending on the country and specific industry, most emerging jobs in a green economy will need a combination of both traditional and green skills. In most cases, it may be more practical to retrain or enhance the skills of workers rather than to replace them with new ones. For low- to middle-skills jobs, traditional skills can be easily supplemented by green skills through on-the-job training programs. This will also largely depend on reorienting the content of the jobs and the way they are performed. For high-skills occupations, more extensive and intensive education and training programs might be required to complement traditional technical skills with a wider and specific set of new green skills (ILO 2011a; UNEP 2011).

Qualifications framework. The existence of a qualifications framework (QF) can facilitate integration of green skills in education and training programs and curricula or modules. A study in 2016 explored how green skills can be integrated in green jobs in Thailand's construction and tourism industries. The main strategy is to retrain workers with a set of well-defined green knowledge and skills developed through the promotion and implementation of green competencies. The initial step is to develop a set of green skills competencies (knowledge and skills) specific to an occupation in the tourism and construction industries. Examples of such occupations are electricians, carpenters, plumbers, and tourism operators (Esposto 2016). Since a QF essentially links qualifications with the actual skills needed in a particular workplace, green skills can be translated into specific competencies that can feed into competency-based training modules or curriculum designs, as well as into assessment and/or certification mechanisms. Integration of green skills in the QF will also allow for workers' mobility across green jobs and sectors.

In a holistic QF that reflects a school level to higher education continuum, generic or basic green skills can already be integrated in the school education values and attitude formation components that include appreciation for and protection of nature and the environment.

Quality assurance mechanism. A rigorous quality assurance (QA) mechanism for skills development covers curricula, programs and institutions, trainers and assessors, and assessment and certification. Integrating green skills standards into QA will not only instruct various education and training institutions but will also guide companies and employers. Quality education and training programs allow transferability of competencies across jobs and enterprises, and, again, mobility of workers across jobs and companies.

In both QF and QA, collaboration among key partners, i.e., government, industries, workers' representatives, and education and training institutions, is critical in identifying the skills needed for renewable energy and in making sure that there is enough provision for education and training. Another crucial partnership is between private companies and their trade associations. In many countries such as Germany and Denmark, governments are working with employers' and workers' associations to adapt or formulate new curricula for renewable energy. In Canada, the Renewable Energy Advisory Committee on Training was created involving the industry, colleges, and the government. In countries where there are existing skills councils, there is already a good venue to start. The Republic of Korea, for instance, added a new sector skills council for renewable energy (ILO and EU 2011a, b).

Synergy across the education sector. Addressing skills gaps while transitioning to a green economy will be greatly aided by forging synergy across education subsectors. In putting emphasis on STEM disciplines, for example, quality at the school level is important. Furthermore, important core attributes such as awareness and attitude related to environmental protection are best developed during basic education.

For technicians and skilled craft workers, TVET apprenticeships with suppliers of renewable energy technologies offer training and exposure and/or emersion. For example, courses on installation and maintenance of wind farms are being offered by vocational training colleges in Spain. In collaboration with employers and unions, a 2-year course to become a geothermal technician is available in British Columbia, Canada. In Brazil, the National Biodiesel Programme provides assistance to rural technicians to help bioenergy crops growers (ILO and EU 2011a, b).

A STEM-focused tertiary education should be adequately provided with incentives (e.g., scholarships and grants, paid apprenticeships) to boost engagement in programs that produce the much-needed researchers and engineers for green jobs, particularly in renewable energy. Courses in areas such as engineering, biosciences, geosciences, agriculture, forestry, and business in universities serve as foundations programs for high-skilled level occupations in renewable energy. Many universities are shaping their programs to accommodate interests in renewable energy. The Oregon Institute of Technology in the US, for example, offers a first degree in renewable energy engineering. An increasing number of universities also offer postgraduate courses on renewable energy. These include the University of Auckland in New Zealand, which offers a postgraduate diploma in geothermal energy technology, and the Faculty of Technology at Makerere University in Uganda, which offers a master's degree program in renewable energy (ILO and EU 2011a, b). Recently, one of the Philippines' state universities launched a professional science master's in renewable energy engineering open to all licensed engineers in the country.

Inclusivity. In pursuing skills development for green jobs, strategies, plans, and programs should be inclusive to ensure equal opportunities for women, youth, and the marginalized. For example, overcoming gender barriers to train and engage women in high-level skills for occupations in renewable energy such as technical and engineering jobs can also help ease the problem of skills shortages. This can be achieved by opening up relevant education and training opportunities, ensuring the quality of programs, making them accessible, and providing guidance and incentives.[6]

---

[6]ILO. Skills and Employability. In Christine Evans-Klock, *Employment for Social Justice and a Fair Globalization. Overview of ILO programmes.* Skills and Employability Department, ILO. http://www.ilo.org/wcmsp5/groups/public/@ed_emp/documents/publication/wcms_140945.pdf.

Women and Renewable Energy

Currently, about 3 billion people still depend on open fires and traditional biomass such as wood and crop waste for cooking and heating. Most of these people live in Asia and Sub-Saharan Africa. Women and children perform most of the work to collect biomass fuels (UNEP 2016b). The availability of various forms of renewable energy and greater energy efficiency not only address climate change, but also create opportunities for energy access to the billions of people who are still not enjoying modern energy services.

For women, energy access improves opportunities by enabling them to engage in more productive activities:

- Access to energy makes routine household work easier and reduces the time taken, allowing women more time to engage in higher income jobs and entrepreneurial activities.
- Access to energy improves basic living conditions for women, including possible access to computers and information, leading to better standard and quality of life for their families.
- Employing women in the electricity or renewable energy sector, for example, allows them to participate directly in the value chain.

4 Way Forward

Green jobs creation is expected to further accelerate in the coming years as international and national initiatives to combat climate change and adopt a more sustainable approach to economic growth intensify. Transition to a sustainable economy has the potential to create green jobs across economic sectors as investments in new technologies, equipment, buildings, and infrastructure continue to increase and serve as key drivers for new employment and an impetus for retraining and transforming existing jobs. This is already happening in both developed and some developing countries.

Transition to green growth brings about skill shortages due to an increasing demand for green occupations. This does not happen in the energy sector alone, in particular renewable energy, but also across other sectors like agriculture, building, and transport. Failing to address skills shortages during the transition to sustainable development will almost certainly derail a country's pursuit of green growth. Therefore, reconsidering and replanning current skills development strategies are imperative to accommodate green growth policies. Initially, for skills development policies targeted at green jobs to be effective, they need to be integrated into national development plans, accompanied by purposive advocacy and adequate awareness campaigns designed to impact targeted trainees and students, as well as the industries and employers.

Orient Education and Skills toward Sustainability

It is increasingly clear that education and skills, attitudes, and behavior are crucial for sustainable and inclusive growth. The 2030 Agenda for Sustainable Development (UN 2015) requires a comprehensive response in terms of human capital development, talent solutions, and education and training. The Global Education Monitoring Report of 2016 (UNESCO 2016) reinforces the role of education in every dimension of sustainable development. Better education leads to greater prosperity, improved agriculture, better health outcomes, less violence, more gender equality, higher social capital, and an improved natural environment. Education and skills also provide the key tools—economic, social, technological, and behavioral—to take on the Sustainable Development Goals (SDGs) and to achieve them.

The manner in which the Sustainable Development Goal (SDG) for education (Goal 4) links with other SDGs is as follows:

Goal 1: Education is critical to lifting people out of poverty.

Goal 2: Education plays a key role in helping people move toward more sustainable farming methods, and in understanding nutrition.

Goal 3: Education can make a critical difference to a range of health issues, including early mortality, reproductive health, spread of disease, healthy lifestyles, and well-being.

Goal 5: Education for women and girls is particularly important to achieve basic literacy, improve participative skills and abilities, and improve life chances.

Goal 6: Education and training increase skills and the capacity to use natural resources more sustainably and can promote hygiene.

Goal 7: Educational programs, particularly nonformal and informal, can promote better energy conservation and uptake of renewable energy sources.

Goal 8: There is a direct link among such areas as economic vitality, entrepreneurship, job market skills, and levels of education.

Goal 9: Education is necessary to develop the skills required to build more resilient infrastructure and more sustainable industrialization.

Goal 10: Where equally accessible, education makes a proven difference in social and economic inequality.

Goal 11: Education can give people the skills to participate in shaping and maintaining more sustainable cities, and to achieve resilience in disaster situations.

Goal 12: Education can make a critical difference in production patterns (e.g., with regard to the circular economy[7]) and in consumer understanding of more sustainably produced goods and prevention of waste.

Goal 13: Education is key to mass understanding of the impact of climate change and to adaptation and mitigation, particularly at the local level.

---

[7]Circular economy aims to eradicate waste—not just from manufacturing processes, as lean management aspires to do, but systematically, throughout the life cycles and uses of products and their components (McKinsey and Company 2017).

Goal 14: Education is important in developing awareness of the marine environment and building proactive consensus regarding wise and sustainable use.

Goal 15: Education and training increase skills and capacity to underpin sustainable livelihoods and to conserve natural resources and biodiversity, particularly in threatened environments.

Goal 16: Social learning is vital to facilitate and ensure participative, inclusive, and just societies, as well as social coherence.

Goal 17: Lifelong learning builds capacity to understand and promote sustainable development policies and practices.

Source: ICSU and ISSC (2015)

Keep up with the Energy Transition: Addressing the Skills Gap

Expansion of renewable energy would appear to have reached a point of inflection in 2015. Global investment in renewable energy expanded significantly, growing by 5%. Renewable energy generation costs continue to fall, particularly in solar photovoltaics, enabling countries to step up investments in renewable energy. Developing economies jumped ahead of developed countries for the first time in 2015 in terms of total new renewable energy investment. The share of global investment accounted for by developing countries rose from 49% in 2014 to 55% in 2015, going ahead of developed economies, with the dollar commitment at $155.9 billion in 2015, compared with an investment of $130.1 billion by developed countries.

Growth in investments in renewable energy has been accompanied by growth in jobs in renewables. IRENA (2016) estimates that global renewable energy employment increased by 5% in 2015 to reach 8.1 million. For emerging economies, there are a number of co-benefits from renewable energy: first, distributed and off-grid renewables are contributing significantly to energy access of deprived households; second, there is a spread of clean energy; third, off-grid renewables are leading to a large number of jobs at all levels of the spectrum including low-skilled and rural employment as well as a spurt of enterprises for the distribution and O&M of off-grid energy systems; fourth, employment of women in renewable energy is found to be greater than in the energy sector as a whole.

As the transition to renewables keeps pace, it is expected that jobs in the sector will continue to grow. It is estimated that doubling the share of renewables in the global energy mix would result in more than 24 million jobs worldwide by 2030 (IRENA 2016). It is clear that this expanding labor requirement in the renewable energy sector will require more investments and planning for appropriate skills and training, support to entrepreneurship development and training in off-grid systems and strengthening firm-level capabilities, and promotion of education and training.

Develop Skills to Support Expansion of Smart Cities

Recent times have seen the sprouting of "smart" cities. This is happening not only in developed countries, but also in developing countries. The rise of the smart city is seen as the direction to ensure the viable and sustainable development of urban centers, given that two-thirds of the global population are expected to inhabit cities by 2045. Smart cities are expected to use technology, but also to provide comprehensive processes and services to improve the quality of life for their citizens. In addition to digital technologies, expertise in accounting, financial management, and governance is required.

A recent report, *Smarter Cities, Simpler Cities*, published by the Association of Chartered Certified Accountants (ACCA 2016), expounds on a crucial theme: It is the people, not just the technology that makes a city "smart." The report highlights the necessity for smart principles to be embodied in the skills of key professionals, as much as in the city infrastructure itself. It is these professionals who will support innovations for bringing sustainable and efficient practices to the traditional functions of a city such as waste management, energy supply, and transport infrastructure. Therefore, planners, city administrators, and municipal workers are expected to play an ever more crucial role in using data insights and applying intelligent systems to help cities to function well and flourish, even as population numbers expand significantly. There is thus a need to invest in the development of skills in professionals to enable them to carry out the functions required to build and maintain the cities of tomorrow.

Respond to Growth in Environmental Goods and Services

There is a clear trend of fast-paced growth in the market for environmental goods and services, which was estimated at $866 billion in 2011 according to Environmental Business International (EBI 2012) and is expected to rise to $1.9 trillion by 2020. While the size of the market is more substantial in developed countries, it is the developing countries that are exhibiting faster growth rates (Table 2). While the biggest markets are concentrated in the US, western Europe, and Japan, the fastest growth rates are found in developing countries in Asia, the Middle East, and Africa, which exhibited growth rates of 9–10% during 2011.

Table 2 Market size and growth of environmental goods and services by country/region

Country/region	Market in $ billion (2011)	% Growth (2011)
United States	311.3	5
Western Europe	256.0	2
Japan	103.3	(1)
Rest of Asia	78.0	9
Latin America	28.5	5
Australia/New Zealand	13.6	2
Central and Eastern Europe	13.7	4
Middle East	17.5	9
Africa	10.3	10

( ) negative
Source EBI (2012)

Environmental services have been estimated by EBI to make up approximately 65% of the environmental industry as a whole. While many environmental services require some environmental goods in their provision, the sale of an environmental product usually involves embedded environmental services content or requires some form of associated installation, maintenance service, and monitoring. For example, in the photovoltaic industry, it is estimated that the rooftop installation cost of photovoltaic modules accounts for 60% of the total cost of purchase. Others have estimated that for every megawatt peak of photovoltaic modules installed, on average 20 manufacturing and 13 installations, maintenance job years will be created. With improved efficiency and massive expansion in production over the last 10 years, the unit cost of photovoltaic modules has been falling. This trend is likely to incentivize greater sales of photovoltaic modules, which in turn would signify a growing demand for related installation and maintenance services.

It is anticipated that the environmental services sector will become increasingly more important in the coming years. Many developing countries, because of their developmental stage, are now beginning to invest more in environmental infrastructure, with stronger regulatory frameworks. These trends are creating new and evolving markets for environmental services and call for investments in developing the skills of professionals rendering environmental services at the higher end of the spectrum but also at the bottom of pyramid markets in rural settings. Environmental services can support the needs of domestic and rural markets but can also tap into the higher order global value chain with adequate capacity and skills.

Improve quality of education and training at all levels. A quality basic education will ensure that a future workforce is equipped with basic learning tools and flexibility for skills development at TVET and higher education levels. A construction worker or electrical technician who had quality training and has enough experience in a certain sector can be easily trained to adapt to green jobs or jobs in the renewable energy sector. For high-skilled workers in green jobs where shortages are more severe, it is worth giving more attention to STEM, strategizing to encourage students to participate in STEM courses. Many institutions are emerging to provide initial or continuing education and training for green jobs. However, quality assurance mechanisms should be in place to maintain standards, including training infrastructure and facilities.

Target girls, females, and disadvantaged. Polices targeted at green jobs should be able to expand the access of women (as well as the youth and the marginalized) to quality education and training at all levels. Their participation in skills development programs relevant to renewable energy should be encouraged through targeted scholarships and apprenticeships, among others. Strengthening women's leadership and participation in sustainable energy solutions and engagement in relevant occupations, particularly in high-level jobs, is critical in the transition to sustainable growth and to end poverty.

Establish strategic partnerships. Forging effective partnerships on all fronts to strengthen education and training, STEM, and R&D in renewable energy will advance skills development goals for green growth. Such partnerships can take the form of international collaborations among institutions such as centers of excellence, universities, and TVET institutions, as well as among government agencies, education institutions, and companies and firms, and employers. Bilateral collaborations among economies can promote skills standardization and mutual recognition. Development partners such as the International Labour Organization (ILO), Asian Development Bank (ADB), and World Bank have been supporting countries in skills development programs and projects. Prudent and efficient use of their assistance will benefit national skills development strategies.

Mobilize and prioritize resources. On top of private or individual willingness to finance training and public provision to fund national skills development agenda, there are many ways to mobilize funds for skills development. Among them is providing incentives to firms that are willing to invest in reskilling and/or upskilling their workers and managers. Another means is the establishment of a national skills fund such as that of Malaysia. And finally, there is assistance provided by development partners. However, the efficient use of resources is as important as the availability of resources. As such, it is important to identify priorities and to adhere to them.

Promote innovations and technologies. Innovations and technologies will be indispensable in pursuing sustainable development. Governments need to establish a conducive environment to encourage engagement in STEM and participation in R&D through policies and incentives channeled to universities, research institutions, technology centers and networks, and private firms, among others. Again, partnerships are critical in the efficient use of resources through sharing and collaboration. Promotion of innovations and technologies will need well-functioning institutions, sound research and education infrastructure, and links between public and private innovation actors and among enterprises committed to R&D.

Strengthen Professionals and Workers at All Levels for Sustainability

Emerging economies need to plan and invest adequately in talent development and skills training in key disciplines and areas that anticipate the needs of a future sustainable society. Table 3 captures a few of the disciplines that need strengthening through education and training.

ADB's Support for Green Growth

Asian regions registered the highest growth rates in the past decade, and if such a pattern continues, Asia's gross domestic product (GDP) is estimated to constitute more than 50% of the global GDP in 2050. Asia's urban population is also expected to almost double (1.6 billion to 3.1 billion). Cities will be the centers of higher education and hubs for innovation and technological development. Buildings and transport systems in urban areas will account for most energy consumption (ADB 2011). This will require establishing and/or partnering with existing centers of excellence dedicated to training in green skills and expanding this capacity.

ADB supports the SDGs. ADB's new upcoming long-term strategy leading to 2030 will describe how ADB aligns with the Sustainable Development Goals (SDGs) and the new global climate agreement. Sustainable development will be the unifying theme for ADB's future initiatives under various sectors. For example, in

Table 3 Professionals and capacities for sustainable economies and societies

Area	Components	Professional or technician responsible
Finance for sustainability	Reducing costs and boosting efficiency, green finance, enabling poorer people to access clean energy through innovative payment systems	Bankers, accountants, financial correspondents, mobile payment operators
Sustainable transport	Awareness of sustainable transport mechanisms, design and planning of bus rapid transport corridors and their management, more efficient engines, hybrids, fuel cells, alternative fuels	Engineers, surveyors, bus operators, manufacturers, urban planners, fare collecting system operators
Green cities development	Green buildings, green spaces, cycling corridors, recycling plants, water-harvesting structures	Green architects, planners, financiers, technicians for green buildings, recycling plants
Innovations in financing for clean energy	Pay-as-you-go utilities and energy services for off-grid customers	Finance professionals, mobile apps and technicians, off-grid energy system managers and O&M technicians
Renewable energy	Distributed and off-grid renewable energy, access to modern energy sources, biomass and biogas for cooking and motive power for agroprocessing and watermills	Solar photovoltaic engineers and technicians, O&M technicians
Energy efficiency	Energy-saving appliances, industrial processes, electrical motors, insulation	Engineers, technicians, construction workers, installation workers
Smart cities	Digitally connected cities with communication infrastructure, connected devices, automated municipal processes. parking solutions, lighting solutions, and Wi-Fi	Providers, manufacturers of sensors, development companies or software providers
Environmental services	Solid waste collection, street and drain cleaning, environmental management plan/compliance/monitoring and auditing, environmental impact assessment, resource management studies, water resource management, environmental management systems/ISO 14001, environmental risk assessment	Environment engineers and technicians, finance specialists, municipal workers, environment economists, energy-efficient systems installers

Source *Compiled by authors*

transport and building, in anticipation of rapid urbanization growth, utilization of energy-efficient and safe urban infrastructure and mass transit will be encouraged and supported. In agriculture, utmost consideration will be given to the protection of soil fertility and water for sustainable food production.

Asia needs to step up further in terms of technology and innovations in terms of energy efficiency, shifting more to renewable energy and providing the necessary skills needed to sustain these moves. As such, in various key sectors that ADB supports, the success of any undertaking depends largely on the quality of human resources. The fundamental strategy is quality education and training at all levels.

Chapter 2
Introduction

Abstract The international agenda of sustainable and inclusive growth is described and analyzed. Key aspects of sustainable development are explored including the Human Development Index, and the Ecological Footprint. The precise meaning of key terms are explained and discussed, namely green jobs, green skills, green growth, and sustainable and inclusive growth concerning economic and social development. The importance of learning providers' conceptualisation of green jobs and green skills, decent work and jobs, generic green skills, United Nations (and the Education for Development community) policy directions for greening, and policy challenges faced by countries are discussed. Sector issues and implications for TVET providers, business and government are analyzed. Recommendations across all four case study countries (India, Indonesia, Sri Lanka, and Viet Nam) are presented, compared, and discussed with particular reference to lessons learnt and emerging best practices. The research design, aims, scope, and research methodology for the research reported on in this book are explained.

Keywords International agenda for sustainable and inclusive growth · Human development index · Ecological footprint · Training pathway · Decent work and decent jobs · Generic green skills

1 The International Agenda of Sustainable and Inclusive Growth

The concept of the green economy has gained widespread acceptance and traction since the global financial crisis of 2008. The idea of mobilizing economies, societies, education, and training for climate change was reaffirmed at the United Nations 21st Conference of Parties in Paris in 2015 (Platform for Advancing Green Human Capital 2016). The United Nations Environment Programme (UNEP) defines a green economy as one that:

> …results in "improved human well-being and social equity, while significantly reducing environmental risks and ecological scarcities" (UNEP 2010). In its simplest expression, a

R. Maclean et al., *Education and Skills for Inclusive Growth, Green Jobs and the Greening of Economies in Asia*, Technical and Vocational Education and Training: Issues, Concerns and Prospects 27, https://doi.org/10.1007/978-981-10-6559-0_2

> green economy is low-carbon, resource efficient, and socially inclusive. In a green economy, growth in income and employment are driven by public and private investments that reduce carbon emissions and pollution, enhance energy and resource efficiency, and prevent the loss of biodiversity and ecosystem services (UNEP 2011: 16).

Along with the widespread acceptance and adoption of the concept of the green economy in policy circles was a strengthening of the international agenda for sustainable and inclusive growth. This agenda has emerged mainly because of the recognition that "creating new wealth through a 'brown economy' model based on fossil fuels has not substantially addressed social marginalization, environmental degradation and resource depletion." In addition, we are still "far from delivering on the Millennium Development Goals by 2015" (UNEP 2011: 17). The United Nations Capital Development Fund (UNCDF 2013) elaborates further that global inequality has been on the rise for some time, and disparities between and within countries are of significant concern in both developing and developed countries. For example, in the three decades since 1985, wage gaps have widened and household income inequality increased in a large majority of Organisation for Economic Co-operation and Development (OECD) countries (OECD, December 2014).

Global inequality is also expressed as differences in national consumption of the biocapacity of the planet, and as the differences in the health of ecosystems. This is of particular concern, considering that since the 1970s, the global annual demand on the natural world has exceeded what the earth can renew. Currently, we require 1.5 planets, which under a business-as-usual (BAU) scenario, is expected to increase to about 2 planets by 2030, and to 2.9 planets by 2050. The ecological footprint of high-income countries dwarfs that of low- and middle-income countries. Examples of ecological footprints in global hectares (gha) by consumption are: Australia, 9.3; United States (US), 8.2; United Arab Emirates, 8.4; Germany, 5.3; Canada, 8.2; United Kingdom (UK), 4.9; The Republic of Korea, 5.7; Malaysia, 3.7; Japan, 5; Brazil, 3.1; The People's Republic of China, 3.4; Indonesia, 1.6; India, 1.2; Kenya, 1.0; Sri Lanka, 1.3; and Viet Nam, 1.7 (Global Footprint Network 2012). For comparison, in 2010, 1.7 gha per person were available. However, this assumes that no land is set aside to meet the needs of other species that consume the same biological material as humans (Global Footprint Network 2014). Ewing et al. (2010) and WWF (2014) demonstrated that no country has achieved a high level of development (as defined by the United Nations Development Programme [UNDP]) within the global biocapacity (Box 1).

Box 1. Human Development Index and Ecological Footprint, 2010

The UNDP defines a high level of development as a Human Development Index (HDI) score of 0.8 or above, while 1.8 gha is the average productive area available for each person on the planet. Countries with an HDI score of 0.8 or higher, and a footprint of 1.8 gha per person or lower, meet two minimum criteria for global sustainable development: a high level of development and an ecological footprint per person that could be globally replicated to a level less than the global biocapacity. Countries that meet both criteria would be shown in the lower right quadrant. Despite growing adoption of sustainable development as an explicit policy goal, no country meets both minimum conditions.

Source Adapted from: Ewing et al. (2010: 21–22)

In addition, the disproportionate demand of the high-income countries leading to the decline of the health of the planet's ecosystems has occurred at the expense of middle- and low-income countries (Box 2, Figs. 1 and 5).

Box 2. Per Capita Footprint of High-Income Countries

High-income countries have historically had the most rapid increase in "per capita footprint." In contrast, middle- and low-income countries demanded less than the average per capita biocapacity available globally until 2006, when middle-income countries exceeded this value.

The Living Planet Index for high-income countries shows an increase of 10% in biodiversity between 1970 and 2010. This is likely to have been due to a combination of factors, not least being that these nations are able to purchase and import resources from lower income countries, thereby simultaneously degrading the biodiversity in those countries while maintaining the remaining biodiversity and ecosystems in their own "backyard." In stark contrast, the index for low-income countries declined by 58%. This trend is potentially catastrophic, for biodiversity as well as for the people living in those countries.

Source Adapted from WWF (2014: 60)

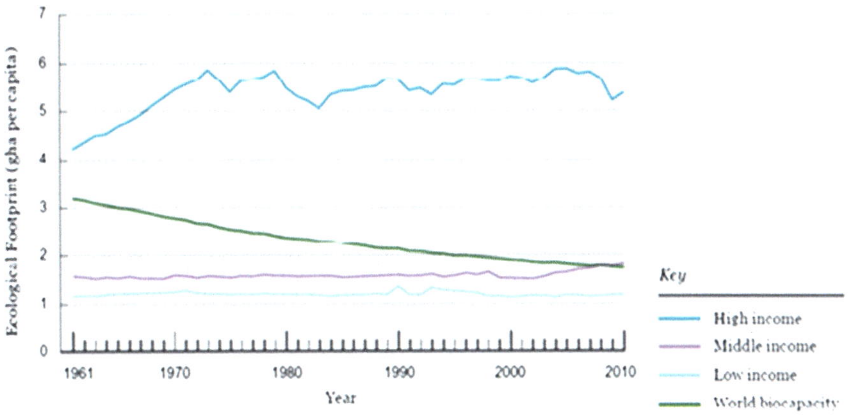

Fig. 1 Changes in the Ecological Footprint per Person in High-, Middle-, and Low-Income Countries, 1961–2010. *Note* The green line represents world average biocapacity in 2010. *Source* Global Footprint Network in WWF (2014).

Inclusive growth has different meanings depending on the national circumstances and level of development, the perspective of the representing body, and other factors. Inclusive growth has both an outcome and a process dimension, and the focus can be on both or either of the two (for an analysis of the concept, see Ranieri and Ramos 2013). It refers to participation and benefit-sharing—participation in the growth process, both in terms of decision-making to organize the growth progression as well as to participate in the growth itself; and benefit-sharing in terms of everyone equitably sharing the benefits of growth.

For some, inclusive growth addresses inequalities only in terms of poverty reduction and income disparities. For others, it includes issues such as access to quality public services including education and health care, political participation, and environmental equality. Inclusive growth also refers to addressing discrimination in gender, age, rural and indigenous populations, persons with disabilities, ethnic and racial minorities, and other marginalized people. In sum, inclusive growth can mean growth that reduces disadvantages and that embraces the non-income dimensions of well-being.

Work on the definition and indicators of inclusive growth are ongoing. The work of the UNDP International Policy Centre for Inclusive Growth starts from the premise that societies based on equality tend to perform better when it comes to economic development. For instance, countries with more equal income distribution are more likely to reduce poverty than countries with unequal income distribution. The OECD held its first Workshop on Inclusive Growth in April 2013[1] with the aim of defining and developing measures for inclusive growth, criteria for policies, and strategies for change. Their vision for inclusive growth is one that combines strong economic growth with improvements in living standards and quality of life outcomes (such as good health, jobs and skills, a clean environment, and community support). However, the OECD warns that, because the rise in inequality is so deeply embedded in our economic structures, it will be a challenge to aim for inclusive growth (OECD 2015: 21).The Asian Development Bank (ADB) states that inclusive growth has become a strategic pillar for guiding its activities. ADB advances a definition of inclusive growth as nondiscriminatory and disadvantage-reducing growth (Klasen 2010).

The combined concept of "sustainable and inclusive growth" is subject to a range of interpretations, with gradual shifts in focus on intersecting sliding scales, and with the three pillars being environment, society, and the economy. Sometimes the term is simply used as a synonym for "the green economy." The European Union (EU) embraces the concept of "sustainable and inclusive growth" as its 2020 strategy, referring to environmental sustainability and growth that includes the disadvantaged and all European countries.

There are clear challenges ahead with regard to the international agenda of "sustainable and inclusive growth" as part of the green economy movement. As Ewing et al. (2010: 21) state:

---

[1]www.oecd.org/inclusive-growth/.

> …even optimistic forecasts are still not sufficient to bring demand within the biological capacity of the Earth. Therefore, relying on a growing level of consumption to attain sustainable well-being for all is unrealistic, especially given the increasing global population. [Placing complete reliance on continued technological improvements] in the future does not represent good planning. Worse, the accumulated ecological debt from decades of ecological overspending is likely to start decreasing the biosphere's regenerative capacity at the same time we are increasing our demands on it. Realizing the "right to develop" of all countries […] requires constructing new development pathways that place much less strain on the global environment than have historically been the case.

It is essential to be clear about the precise meaning of key terms such as green jobs, green skills, and inclusive growth adopted in this book. In this research study, **inclusive growth** is taken to mean growth that reaches and benefits all members of a community and society in concrete ways. **Green growth** refers to patterns of decent work and the production of goods and services that are environmentally friendly and that minimize destruction of the environment. Green jobs is defined in Box 3.

A comprehensive review of the research and related literature concerning education and skills for inclusive growth and green jobs can be found in the regional summary of this book.

Greening does not imply a new sector and does not necessarily mean the creation of new jobs but it suggests a restructuring of the employment market in ways such that some new employment opportunities would be created. These could include manufacturing of solar panels and pollution-control devices. In addition, some jobs would be substituted through change, for example, a shift from fossil fuels to renewables or landfilling and waste incineration to recycling. Certain jobs could also be eliminated with the discontinuation and banning of plastic bag production and use. Many jobs would be transformed and redefined such as those of plumbers, electricians, metal workers, and construction workers, as day-to-day skills sets, work methods, and profiles are greened (UNEP 2008).

Green jobs have implications across sectors and are not limited to the familiar renewable energy and energy efficiency sectors. Different shades of green jobs are manifested in infrastructure development and construction, power, transportation, travel and tourism, agriculture, and forestry.

Box 3. Definition of Green Jobs Adopted in this Research Study

Jobs that reduce the environmental impact of enterprises and economic sectors, ultimately to levels that are sustainable, are termed green jobs. This comprises work in agriculture, industry, services, and administration that contributes to preserving or restoring the quality of the environment while also meeting requirements of decent work—adequate wages, safe conditions, workers' rights, social dialogue, and social protection. This definition is in keeping with that adopted by the International Labour Organization and relevant others (ILO/UNEP/IOE/ITUC 2008).

The Pro Enviro (2008) report found that moving to a low-carbon and resource-efficient economy (LCREE) required a fundamental transition in behavior and the application of new skills and knowledge. Understanding and awareness were also identified as a crucial issue. The report also found that the interchangeable use of terms such as "sustainable development," "green," "eco," and "environmental" caused confusion. This was the case with the term "green jobs" in the context where sustainable employers considered all jobs to be green. The Pro Enviro report further identified a range of generic (cross-sector) and sector-specific skills as priorities. Many of the identified skills already existed, had increased demand, or needed application in new situations. The report concluded that the integration of LCREE skills into all training undertaken by companies is the key to mainstreaming sustainable understanding, knowledge, skills, and thinking.

Her Majesty's Government (2011) report, *Skills for a Green Economy*, states that companies should be thinking about generic skills required to use resources efficiently and sustainably. Additionally, the report notes that there is demand for a workforce with generic green skills across all types of firms and sectors, and that some traditional skills require a distinctive green economy awareness and understanding. In addition, learning providers have shown interest in modifying curricula to accommodate new learning demands (Box 4).

The European Center for the Development of Vocational Training (CEDEFOP) (2010) study on green skills also argues that a balance of generic and generic green skills is needed for low-carbon jobs. Along with topping up existing job-related skills, this is much more important to developing a low-carbon economy than are more specialized, green skills. The CEDEFOP study draws parallels between green skills and information technology (IT) skills, which were once considered solely specialist skills but which are now pervasive through all fields of work and life. Research by the United Nations Educational, Scientific and Cultural Organization (UNESCO), and particularly by the UNESCO International Centre for Technical and Vocational Education and Training (UNESCO-UNEVOC) highlights the lack of understanding of the concept of green skills and green jobs in both the organized and the unorganized sectors. Majumdar (UNESCO-UNEVOC 2011) argues for a whole institutional approach toward TVET greening that focuses on curriculum, campus, community, research, and culture.

Box 4. Role of Learning Providers' Research Study

Learning providers express enthusiasm and willingness to change the content of curricula to meet new demands. However, actions have been rather piecemeal and reactive so far, although less so among those providing tuition for new occupations such as energy auditors and solar photovoltaic installers. Some learning providers gave cautious responses because of concerns about insufficient demand, and others said that they would probably wait to see how the market developed before investing. Most learning providers do not yet include green skills in their learning strategy.

Source CEDEFOP 2012: 9

A recent study by Eccles et al. (2012) suggests that companies that are early adopters of sustainability practices are significantly more likely to establish formal stakeholder engagement processes compared with companies with a low focus on sustainability. Additionally, an understanding of standards and legislation, and the ability to implement these, and skills in auditing and assessment, are important for sustainability and green business practices.

Demand for skills specific to expanding industries such as renewable energy is growing, while demand for skills in fossil fuel-based industries is declining. There is also a demand for reskilling or upskilling jobs that are transforming due to the changing nature of businesses, regulations, and compliance toward energy-efficient technologies and practices.

According to the OECD, there is considerable evidence that skills shortages may be impeding the transition to green growth in sectors such as energy-efficient construction and retrofitting, renewable energy, energy efficiency, and environmental services. Skills shortages already appear to be impeding the greening of growth. In the People's Republic of China (the PRC), as well as in India, rural electrification programs are suffering from a lack of skilled workers. Reasons for these shortages are the poor reputation and limited attractiveness of sectors such as waste management, and a limited number of teachers and trainers in environment-related services. Countries need to establish pathways for training, for skills development, and for advanced knowledge-building between skills development and TVET and higher education institutions (Box 5).

Box 5. Training Pathways—International Examples

According to the Slingenberg et al. (2008), a number of countries have developed green job profiles to supplement career advice such as the Bureau of Labor Statistics within the United States Department of Labor and the United Kingdom National Career Service, which can draw on consistent systems of classification for wages and qualifications. These skills profiles provide an overview of the different phases in a sector, such as the design, manufacturing, installation, and maintenance phases and identifying the associated jobs and duties for each. The skills profiles also identify the education and training pathways into the careers and salary ranges. The National Association of Regional Councils in the United States has similar green job profiles that provide career ladder information and list specific training institutions delivering relevant vocational training including prerequisites. International industry associations and professional bodies have also developed green jobs profiles. The Australian Power Institute, for example, and some companies such as ENERCON identify key duties, responsibilities, and education and training pathways.

Source Slingenberg et al. (2008)

2 Arguments Concerning the Role of Increased Employment and Productivity in Promoting Sustainable and Inclusive Growth

The prevailing view used to be that economic growth trickles down in society to lift populations out of poverty. Kuznets's hypothesis (1955) (Kanbur in Ranieri and Ramos 2013: 2) states that in the early stages of development, growth produces (and exploits) inequality, and as per capita income rises there comes a turning point after which inequality declines. It was argued that, at first, growth not only produces but necessitates income inequality. However, as Ranieri and Ramos (2013: 2) demonstrate, there has never been "consummate empirical evidence" of that being the case. In advanced economies, it can be seen that real per capita income grew steadily from 1950 to the late 2000s, but inequality also grew. In some developing countries, average income steadily grew, but in others, fast growth was accompanied by worsening inequality without the growth process reversing that trend. Instead, poverty rates persisted at high levels or even increased (Ranieri and Ramos 2013: 2).

The phenomenon of jobless growth has resulted in a deepening of existing inequalities and social exclusion, and the existence of the phenomenon of the working poor, where the majority of people—while working—remain poor and deprived (UNDP, UN Department of Economic and Social Affairs, and International Labour Organization (ILO), 2012: 4). Low growth is predicted for many large economies, and job creation will remain a challenge for some time. Moreover, the poor living in both rich and poor countries are most vulnerable to environmental degradation and climate change (OECD 2010a, b). Since low-income households spend a significant share of income on food and energy-related items, they are also more exposed to the impact of wasteful practices, overconsumption, and environmental degradation, including soil degradation, deforestation, overfishing, and fresh water shortage (ILO 2012a, b: 1).

In addition, globally, a link has been shown to exist between environmental pollution and social deprivation, with poorer communities being more exposed to environmental hazards. For example, the Commission of the European Communities found that Europe's poorest communities suffer the worst air quality, are more likely to be located near industrial sites, and are more exposed to the risk of flooding. In one study cited, people in deprived communities were exposed to 41% higher than average concentrations of nitrogen dioxide (NO_2), and in another it was found that half of municipal waste incinerators in the United Kingdom were located in the poorest 10% of communities (Commission of the European Communities 2005: 16). Other studies throughout Europe, the United States (e.g., Environmental Health News 2012), and in developing countries confirm this link.

ILO points to the need to pay attention to the quality of jobs, captured in the concept of "decent work," to help the working poor break out of the "vicious circle" of exclusion from economic growth (ILO 2011a: 21–23). Moreover, it is now acknowledged that inequality matters, particularly for the long-term sustainability of growth (Ranieri and Ramos 2013: 4). Evidence is mounting that the business-as-usual (BAU) model bears high environmental costs, in particular as a consequence of climate change, as well as social costs, to a degree that peace and security are put at risk within nations and globally. A shortage of productive employment is considered to be a threat to social cohesion, since it produces social resentment and contributes to social and political instabilities. The aim is to facilitate access to employment for under-represented groups such as youth, older workers, women, and migrants. This requires "decent" work, that is, jobs that enable people to avoid and escape poverty. Beyond its fundamental economic role, work is acknowledged as a means to build identities and social cohesion, to participate in and contribute to society, to gain self-esteem, and to feel a sense of belonging to a community.

The current growth model is also economically unsustainable. Initial findings from the Economics of Ecosystems and Biodiversity series point out, for instance, that annual loss in biodiversity and ecosystem services due to deforestation and forest degradation is equivalent to $25 trillion. In addition, higher concentrations of greenhouse gases (GHGs) in the atmosphere will entail considerable economic costs, notably in terms of reduced productivity. Projections indicate that, under a BAU scenario, productivity levels will be reduced by 2.4% in 2030 and by 7.2% in 2050 (ILO 2012a, b: 4) (Fig. 2).

Modelling by the United Nations Environment Programme (UNEP) (2011: 518–521) demonstrates that an overall decoupling of natural resource uses from economic growth is possible and that it can result in positive trends for employment, gross domestic product (GDP), poverty, nutrition, water stress, footprint, and biocapacity. UNEP states that, depending on the investment simulated, and its timing, the global total net direct employment in green sectors may decline in the short term (primarily due to a decline in fishery and forestry sector employment) to converge on or rise above BAU employment in the medium to long run.

ILO, in collaboration with other agencies, has produced a significant body of evidence that shows that employment and decent work can facilitate the shift toward sustainable and inclusive growth. ILO (2012) emphasizes combining environmental policy with labor policy to bring about a double dividend of increases in employment and improvements to the environment. They argue that even immediate positive socioeconomic benefits are possible if environmental policies are accompanied by appropriate measures that mitigate possible negative consequences.

ILO cites environmental tax reform as the most prominent example. ILO modelling shows that if an eco-tax equivalent to 1% of GDP had been introduced in 2012, and labor taxes had been simultaneously reduced by the same amount, multifactor productivity would be 1.5% higher in 2020 compared with the case in which green taxes are not used to support employment. By 2050, multifactor

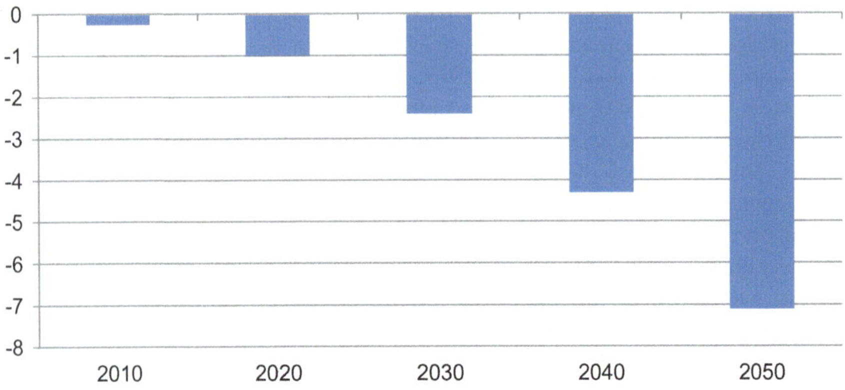

Fig. 2 Productivity Loss from Further Increases in Greenhouse Gases (%). *Note* The baseline scenario assumes that environmental damage remains at the level of the base year (2000). *Source* Bridji et al. in ILO (2012: 4)

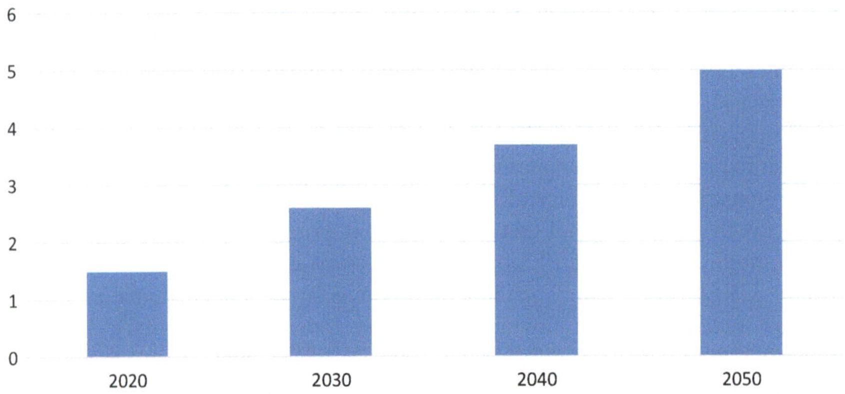

Fig. 3 Global Productivity Effects of Using Green Tax Revenues to Support Jobs (%). *Note* The figure shows the increase in multifactor productivity that is estimated to arise as a result of the use of revenues from green taxes (equivalent to 1% of gross domestic product) entirely to reduce labor taxes. *Source* Bridji et al. in ILO (2012: 168)

productivity would be 5% higher (Fig. 3). The idea is that lower labor taxes boost employment, in turn stimulating potential output and creating new investment opportunities. To amplify the positive effect, ILO suggests using tax revenues for skills development to meet the current skills shortage of green skills. (For more discussion and an overview of the empirical evidence, see ILO 2011b: 69–74.)

The concept of economic growth has lost its primacy in so far as it is now recognized that sustainable development has to build on social inclusion and preservation of the environment.

3 Conceptualization of "Green Jobs" and "Green Skills"

3.1 Green Jobs

The European Commission (2013: 7) has identified two approaches to defining green jobs that have emerged from the policy and research literature: the "eco-industry" approach, in which "jobs are green by nature of activity," and the "transformation" approach, in which "all jobs are greening." UNEP et al.'s definition sits along the lines of the eco-industry approach, referring to specific activities and sectors that contribute to improving the environment. They define green jobs as

> …work in agricultural, manufacturing, research and development, administrative, and service activities that contributes substantially to preserving or restoring environmental quality. Specifically, but not exclusively, this includes jobs that help to protect ecosystems and biodiversity; reduce energy, materials, and water consumption through high-efficiency strategies; decarbonize the economy; and minimize or altogether avoid generation of all forms of waste and pollution (UNEP, ILO, International Organisation of Employers (IOE), and International Trade Union Confederation (ITUC) 2008: 3).

An important element of this definition of green jobs is that the jobs need to be decent work, that is, productive, providing adequate income and social protection, safe working conditions, job security, reasonable career prospects, a respect for the rights of workers, and giving workers a say in decisions that will affect their lives (UNEP, ILO, IOE, and ITUC 2008: 3).

A later definition of ILO suggests that there is room for a job to be included in the definition of green jobs as long as the work is designed to reduce the environmental impact that the work would have under BAU conditions, and as long as the work is decent. They say that jobs are green jobs when they

> …reduce the environmental impact of enterprises and economic sectors, ultimately to levels that are sustainable. Specifically, but not exclusively, this includes jobs that help to protect ecosystems and biodiversity; reduce energy, materials, and water consumption through high-efficiency strategies; decarbonize the economy; and minimize or altogether avoid generation of all forms of waste and pollution (Keivani et al. 2010: ii).

Some jobs in high carbon-emitting industries could be included such as steelworkers producing wings for wind turbines. On the other hand, there are jobs in sectors that generally are understood to be part of the green industry but, because of the processes and practices they use, cannot be considered green jobs. Figure 4 visualizes the spectrum of green and decent work (e.g., UNEP et al. 2008: 3; ILO 2012a, b: 7).

Activity-based definitions of green jobs are likely to focus on a too-narrow area of the economy (European Commission 2013: 8). The "transformation" approach on the other hand considers cross-sectoral impact and restructuring effects on the resource- and energy-intensive sectors. As the European Commission (EC) indicates, the transition to a greener economy is a dynamic process of transformation likely to affect the quantity and nature of work and skills needs across all sectors

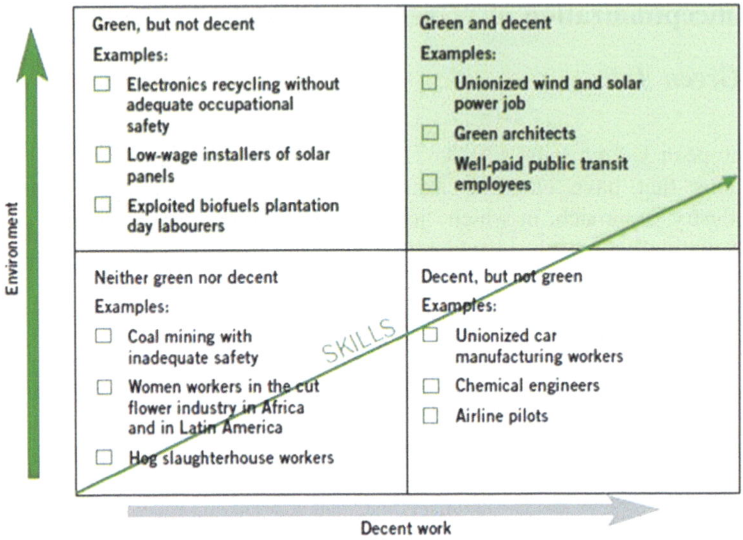

Fig. 4 Green and Decent Jobs—A Schematic Overview. *Source* ILO (2011a: 23)

and occupations (EC 2013: 7). As greener economies develop, the dividing line between "green" and "other" work can be expected to shift, so we can think of "shades of green" when it comes to different jobs. A set of broader definitions have therefore emerged, placing "green jobs" on a spectrum. Jobs or economic activity in this sense are seen as "green" as a consequence of the transformative process of "greening." To this end, the EC defines green jobs as:

> … covering all jobs that depend on the environment or are created, substituted, or redefined (in terms of skills sets, work methods, profiles greened, etc.) in the transition process towards a greener economy (EC 2013: 8).

The EC states that, although the two approaches are conceptually distinct, they are not mutually exclusive. The "eco-industry" approach allows us to identify a core subset of activities that are affected by environmental and low-carbon drivers. These are the kinds of jobs created through new technologies such as renewables, and the redefinition of existing jobs in eco-industries such as agricultural workers who are increasingly required to manage farmland habitats and features, or spatial planners who need to integrate the needs of habitats and species with those of human land uses (EC 2013: 8).

Countries have been developing their own national definitions of green jobs. The report of the OECD (2012: 117–119) provides a list of these. ILO is working on a standardized definition that can be applied by countries in all regions and at all stages of economic and social development (Stoevska in ILO 2012a, b: 7). Stoevska and Hunter (2012, Oct), labor statisticians for ILO, have developed a schematic classification where it is possible to identify which kinds of jobs are excluded or included in the definition for statistical purposes and to allow for comparisons.

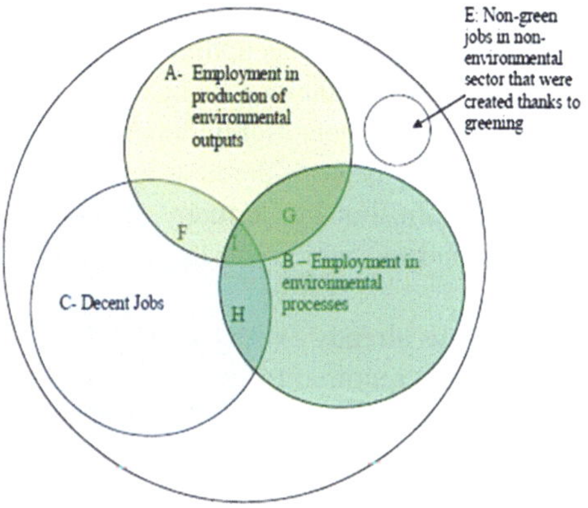

A: Employment in green industries
B: Green occupations
C: Decent jobs
F: Decent nongreen jobs in green industries
G: Nondecent green jobs in green industries
H: Decent green jobs in nongreen industries
I: Decent green jobs in green industries (including green output from nongreen industries)
Group “E” includes, for example, steelworkers producing wings for wind turbines.
Green jobs, according to the UNEP et al. (2008) definition, are those in segments F + H+ I

Fig. 5 Schematic Relationships among Total Employment, Green Jobs, and Decent Work. *Source* Adapted from ILO (2012a b: 7), Stoevska and Hunter (2012, Oct.: 16)

The core elements are jobs defined by environmental outputs and jobs defined by environmental processes (Fig. 5).

3.2 Green Skills

Conceptualizations of green skills range from the narrow technical skills required for very specific jobs to the broadest set of soft skills (ILO 2011a: 103). UNEP et al. highlight that employment and skills needs are affected in at least four ways as an economy is oriented toward greater sustainability:

In some cases, additional jobs will be created—as in the manufacturing of pollution-control devices added to existing production equipment.

Some employment will be substituted—as in shifting from fossil fuels to renewables, or from truck manufacturing to rail car manufacturing, or from land-filling and waste incineration to recycling.

Certain jobs may be eliminated without direct replacement—as when packaging materials are discouraged or banned and their production is discontinued.

It appears that many existing jobs (especially those such as plumbers, electricians, metal workers, and construction workers) will be transformed and redefined as day-to-day skills sets, work methods, and profiles are greened. (Or, for example, gas fitters move toward installing gas combined heat and power instead of traditional systems, or builders develop skills in retrofitting, EC 2013: 8, added by this author) (UNEP, ILO, IOE, and ITUC, 2008: 43).

The new jobs in the green economy are likely to be medium- and high-skilled jobs (EC 2013: 22). ILO (2011a) summarizes that there is a tendency for emerging occupations to require higher level qualifications, while low and medium skill-level jobs will experience changes. Many more jobs in existing occupations face skills changes than there will be jobs in newly emerging occupations. Whether a job is new or simply an existing job but with some new elements highly depends on the country context. For example, an energy auditor may be considered as a new green occupation in, for example, Indonesia or Viet Nam, but in Germany, it could be seen merely as a shift in the competences of an auditor, which is a long-established occupation (CEDEFOP 2010: 13–14).

Fundamental skills for the majority of green jobs already exist (CEDEFOP 2010: 8–9; Bird and Lawton 2009). The level of retraining required for workers to convert to an occupation in a greener industry may be less than expected. Rather than full retraining, it is more likely that workers may need some additional training to learn about new green concepts and practices, upskilling or adding to existing core skills. For example, workers with experience in shipbuilding and in the oil and gas sector are sought after in the wind-turbine industry for their skills in welding, surface treatment, and outfitting. Internationally, for the transition to a green economy, the skills that are in demand and that are missing are those that are in demand and missing in the economy in general: management and leadership skills; and technical, job-specific skills, including those in science, technology, engineering, and mathematics (Bird and Lawton 2009: 8; CEDEFOP 2010: 8).

According to other experts, specific skills profiles in the green economy exist. Examples of these are knowledge of sustainable materials, relevant traditional skills, carbon footprinting skills, environmental impact assessment skills (flora, fauna), and a good understanding of the "sound" sciences (CEDEFOP 2009: 26). Some core skills that apply in one way or another are those in improving energy and resource efficiency; complying with environmental legislation; reducing environmental pollution and waste; and adopting, adapting, implementing, and maintaining skills (for a list of core skills, see ILO 2011a: 107). Generic skills that are needed include strategic and leadership skills, adaptability and transferability skills, systems analysis, design, holistic approach, risk analysis, coordination skills, and entrepreneurship. The complexity of these skills increase with the complexity of the jobs (CEDEFOP 2009: 26).

Appendix A lists some generic green skills.

4 Background and Purpose

The prominence of Asian economies and corporations, together with globalization and technological innovation, are leading to long-term worldwide changes in trade, business, and labor markets. There is a rebalancing of power and influence, centered on the Asia and Pacific region.

The ADB report, *Asia 2050: Realizing the Asian Century (2011)*, indicates that Asia could account for 51% of world GDP by 2050 (compared with 27% in 2010) if it continues to grow at the same pace. This represents a sixfold increase in per capita income, equivalent to European levels of today. Seven economies—the PRC, India, Indonesia, Japan, the Republic of Korea, Malaysia, and Thailand—are projected to account for 87% of GDP growth in Asia and almost 55% of global GDP growth between 2010 and 2050.

Trade within Asia is growing more rapidly than trade between Asia and the rest of the world, potentially leading to a deeper level of regional integration and cooperation, which is a major avenue for regional development. These developments for greater prosperity need to be maintained through inclusive and sustainable growth.

Some Asian countries have pursued policies of industrial upgrading and diversification toward more sophisticated products that add value and increase labor productivity. Others have pursued service-led growth. Countries in Asia need to align skills and training policies with economic and industrial policies to sustain growth, productivity, and competitiveness. Asian countries need to address problems of unemployment and underemployment, particularly among youth and the marginalized. A reengineered, modernized, and innovative skills development system, contributing to employability and sustainable livelihoods, is called for.

With most Asian countries now achieving good progress toward universal primary education, the priority is shifting to secondary and postsecondary education. Countries need to build strong and diversified skills training and higher education systems to supply the expanding needs of employers. In this regard, skills development for employability, particularly through TVET, is viewed as being particularly important (Maclean et al. 2012).

Ensuring that the resultant economic growth is distributed fairly, and does not damage the resource base upon which future growth depends, requires policies for inclusive and sustainable growth.

The transition to green growth for a carbon-constrained future currently focuses on matters such as efficient use of energy; consideration of secondary energy inputs (e.g., from fuel used for transportation of goods and services); greater use of renewable energy and associated investment in technology development; waste reduction leading to lower pollutant emissions; production processes that conserve, recycle, and reuse natural resources; and an understanding that environmental regulations, standards, and economic instruments are not a hindrance to production, but are essential for achieving inclusive and sustainable growth (Maclean et al. 2014).

Climate change is generating economic and environmental dislocations, and these pressures are set to increase in the coming years. These pressures also bring opportunities in the Asia and Pacific region through the necessity to develop new technologies for adjusting to a carbon-constrained future. This will see a reengineering of established production techniques and will increase demand for climate-compatible goods and services. Those places that best anticipate and respond to these needs will be positioned for significant growth in the years ahead.

Regarding inclusive growth and green jobs in Asia, a number of Asian countries have reached middle-income status (ADB 2011). These countries need to improve competitiveness to escape the "middle-income trap." This requires strengthening skills for new technologies and innovation.[2] Policy makers are acknowledging the importance of anticipating and building capacity for future needs (ILO 2010). However, the adoption of green growth policies is likely to require new skills to react to the needs of the labor market. Several governments are building policy foundations toward green growth and the need to prepare for the growth of green jobs and green skills through responsive education and training strategies (ADB 2012).

The ADB *Education Sector Operations Plan* (ADB 2010) seeks to align ADB support in education with the needs and priorities of its developing member countries (DMCs) and strengthen the quality, inclusiveness, and relevance of education. The plan emphasizes that ADB will support DMCs to diversify and expand higher education and TVET to match labor market needs.

To maintain growth rates in Asia, important concerns include increasing employment, facilitating appropriate skills provision for global competitiveness, and reducing inequality. A key factor affecting rising inequality is rapid globalization, which raises the need for new and appropriate skills (World Bank 2010). Economies need skills to serve knowledge-based industries, provide high-level skills to support technology absorption and further innovation, and broaden opportunities for skills development in new and emerging technology areas. Enhancing opportunities for skills training, particularly for poor and middle-income groups, women, and disadvantaged students, in line with the needs of fast-growing industries, will bring employment benefits and higher incomes, and will act as an important contributor to inclusive growth. Youth unemployment, underemployment, and provision of training in out-of-date, irrelevant skills are urgent issues to be addressed.

Maintaining economic growth sustainably is a policy challenge that Asia has begun to address. Awareness is increasing that green growth approaches are economically and politically feasible and profitable. A number of studies estimate that by 2030, 100 million green jobs[3] will be available worldwide (ILO 2008)—about 2% of the global workforce. Of these, 50 million green jobs are anticipated in Asia. Green stimulus packages after the financial crisis of 2008 have led to increased public investment in green infrastructure, for example public transport, low-carbon energy production, smart electricity grids, energy-efficient buildings, and water and sanitation infrastructure (Anbumozhi and Bauer 2010). Examples of green jobs

[2]Analysts have described the "middle-income trap" as one where developing economies are not able to move beyond growth based on low-cost labor advantages in manufacturing for exports to growth that is capital and skill intensive, supports expanding domestic markets and the services sector, and brings greater value addition.

[3]Green jobs contribute to preserving or restoring the quality of the environment and can be in diverse sectors such as energy supply, recycling, agriculture, construction, and transportation. They reduce the environmental footprint of economic activity and help cut consumption of energy, raw materials, and water through high-efficiency strategies to decarbonize the economy and reduce GHG emissions, to minimize or avoid altogether all forms of waste and pollution, and to protect and restore ecosystems and biodiversity (UNEP).

include manufacturing fuel-efficient cars, solar panels, green-building construction work, renewable energy products, and green professionals such as energy auditors (Box 6).

Care needs to be taken to ensure that green jobs provide decent work, with adequate wages, safe working conditions, job security, career prospects, and worker rights.

Box 6. The United Nations and Policy Directions on Greening
Jobs that reduce the environmental impact of enterprises and economic sectors, ultimately to levels that are sustainable, are termed green jobs. This comprises work in agriculture, industry, services, and administration that contributes to preserving or restoring the quality of the environment while also meeting the requirements of decent work—adequate wages, safe conditions, workers' rights, social dialogue, and social protection. This is in keeping with the definition adopted by agencies such as ILO and relevant others working in this area (ILO/UNEP/IOE/ITUC 2008).

Education and training need to meet the requirements of labor markets including those for green jobs. Enterprise surveys by the World Bank show that 50% of firms surveyed since 2000 in East Asia, Southeast Asia, and the Pacific are concerned about the quality of workers' education and skills, compared with 25% of firms surveyed in member countries of OECD (World Bank 2010). The survey found unemployment of graduates and fairly long time intervals to fill professional vacancies. This indicates that education and training systems are mismatched with the needs of the high-growth sectors. Skills training initiatives need to be supplemented with effective skills forecasting models that are responsive to industry plans and requirements.

Countries are evaluating the costs of adopting green growth measures, and industries are being encouraged to move to green production systems through incentives or regulatory requirements. Training is required in new and redefined skills to prepare the workforce for a green economy. New green jobs will become available, and job substitution will result from the greening of corporations and the shift from fossil fuels to renewable sources (UNEP 2008). Education and training institutions do not yet have effective response strategies for green jobs. The ILO Global Report (2011) concludes that in many developing countries governments and formal TVET have not yet responded to skills challenges associated with environmental issues and climate change. This is mainly due to the absence of well-developed mechanisms that help TVET to adjust to the changing needs of the labor market. Often companies account for the biggest share of skills provision, and nongovernment organizations (NGOs) are playing a key role in nonformal skills provision for green jobs. Training course offerings for green skills need to be developed in sectors that are key drivers of growth and employment. Green industries in Asia and the Pacific are currently fragmented. University offerings lack cross-disciplinary breadth and the faculty to train future workers for the technical, economic, social, and managerial challenges in green industries.

Industry and business have initiated steps to train workers for green processes and products. However, governments, industries, and educational institutions have not yet developed coordinated actions to prepare the workforce in green skills. As countries in Asia and the Pacific put in place national policies and strategies for adapting to a carbon-constrained future, the capacities needed have not been adequately analyzed. Regional entities, such as the Association of Southeast Asian Nations, are supportive of bringing together business leaders in green industries. International commitment to clean and green growth requires green professionals in areas such as assessment, reporting, and certification. Currently they are primarily from developed countries.

The ILO (2011a, b) Skills for Employment Policy Brief identifies a number of challenges that countries face. These are summarized as follows:

Challenge 1: Skills shortages are hampering the transition to greener economies.

A lack of the skills needed to meet the requirements of changing and newly emerging occupations impedes green investment and hinders green economic development. Many countries lack teachers and trainers in environmental awareness and specialist areas such as renewable energy.

Challenge 2: Skills and environmental policies need to come together.

Most countries have drawn up environmental policies, but few have put in place the skills development strategies needed to implement these. Without coherence between skills and environmental policies, skills bottlenecks may impede the successful transition to greener production and consumption.

Challenge 3: Green structural change will be profound in certain sectors.

High carbon-emitting sectors are most likely to be negatively affected by green structural change. These include mining, fossil fuel-based energy generation, manufacturing, forestry, and agriculture. Workers moving out of declining sectors into growing ones will require retraining. The role of employment services in matching skills and jobs and in retraining workers and job seekers is thus crucial.

Challenge 4: Occupations will change at different rates and in different ways as economies go greener.

Not all greening will involve a fundamental change in occupations. Some occupations will not change. There will be far more established occupations requiring skills upgrades than new occupations. Where new occupations are created they often call for higher level qualifications, either because of their dependence on new technologies or because they require sophisticated skills.

Challenge 5: Skills required in a greener economy need to be identified.

Ways of classifying and measuring green jobs are only now being refined. In their absence, most countries have relied on qualitative data gathered through enterprise surveys, occupational research, or consultation with experts. There is thus a need to develop complementary quantitative methods to identify the specific occupational needs of a greening economy.

Challenge 6: Appropriate training needs to be put in place promptly.

Training systems need to respond quickly to demand. This is a particularly challenging requirement, given that the updating of courses usually takes time. Government-sponsored training programs have proved valuable, notably where they take advantage of formal education and training systems with well-established flows of information between industries and training institutions.

5 Research Design

5.1 Aims

The key objective of this ADB research study is to increase the knowledge and capacity of the four DMCs being examined (India, Indonesia, Sri Lanka, and Viet Nam) to match education and skills training to jobs, including for green occupations. The project aims to assist these four DMCs put in place timely and effective policies and strategies for skills development. It seeks to facilitate dialogue and networking among the public sector, business and sector councils, industry and occupational associations, and employee associations to assess gaps in policy and practice for the development of skills and to anticipate future needs. The research study addresses both technology-oriented high-end skills as well as middle- and lower end skills to ensure that issues of inclusiveness are addressed in the context of rapidly growing economic sectors, including greening sectors.

In terms of the governance of the four countries there are some important similarities and differences: India, with a population of 1.27 billion, is a democracy involving a union of 29 states and 7 territories; Indonesia, with a population of 242 million, is a presidential representative democratic republic with 34 provinces; Sri Lanka, with 20.5 million, is a presidential representative democratic republic; and Viet Nam, with a population of 93.4 million, is a single-party socialist republic with a centrally planned and controlled economy.

This ADB-Education University of Hong Kong (EdUHK) research study seeks to lead to improved knowledge and enhanced capacity in the DMCs to create an adequate skills base to support growth and employment and to serve green jobs in key industries. Education and training institutions should become aware of best practices for curricula and training approaches to respond to skills needs in key industries. The study addresses industries with potential for high employment, growth and greening, and those that face changes due to a carbon-constrained future. Selected industries have been examined, these being building and/or construction, transport, energy, hospitality, and one other important industry specific to each country's context. The study considers two sectors in detail in each country, not necessarily the same ones for each country, and the other three sectors are covered but in less detail to allow comparisons regionally.

5.2 *Project Scope*

For each of the four participating countries the scope of this study is to

- identify mismatches and gaps between training and employers' needs in green jobs;
- identify opportunities for strengthening green skills training as a means of reducing unemployment for all, including youth and women;
- provide case studies demonstrating good practices; and
- recommend policy and strategy options for skills development, with a focus on green jobs.

5.3 *Methodology*

Data for each of the four countries have been collected through

- surveys (100+ TVET providers and 100+ employers) in each country;
- interviews (50+ interviews in each country with representatives from government, NGOs, the private sector, and TVET providers);
- roundtable meetings (bringing together up to 20 stakeholders in each country for discussion sessions);
- inputs from the three international and 12 national consultants; and
- review of the research and related literature.

For information on the research instruments used see Appendix 3 for the Employers' Survey Instrument; Appendix D for the TVET Providers' Survey Instrument; Appendix E for the Employers' Interview Protocol; and Appendix 6 for the TVET Providers' Interview Protocol.

5.4 *Limitations*

This research study cannot meaningfully speak about "the economy" in India, Indonesia, Sri Lanka, or Viet Nam in overall, general terms, since there are, in a sense, many economies (or subeconomies) in each of these countries: the formal as well as the nonformal economy, the national economy, the various state economies, and the urban and rural economies.

This study focuses mainly on what is happening in the formal TVET sector, and some sectors of the formal economy, in India, Indonesia, Sri Lanka, and Viet Nam. However, the unorganized sector is very large, important, and influential in each of these countries, since it is reliably estimated that over two-thirds of economic activity occurs in the informal sector. This research also has not included such important sectors as retail and agriculture, although a large portion of the population in each of the four countries works in these sectors.

Gathering reliable, representative statistics from government sources is a major challenge, because of the many shortcomings in the coverage, accuracy, and quality of data collected. It is not possible to generalize from government statistics for the whole workforce, since available data refer mainly to the formal sector of the economy. The identified skills are only indicative and are not comprehensive. The skills could not be fully mapped, with occupational standards being developed by the sector skills.

In each country, there were approximately 100 respondents to the survey questionnaire filled in by TVET providers, and a similar number filled in the survey of employers. In addition, in each country approximately 50 interviews were undertaken with employers, employers' associations, TVET providers, government representatives, and NGOs. Given the relatively small numbers, the reporting of the survey and interview data therefore needs to be treated with caution. It should not be assumed that the survey and interview findings are indicative for the countries as a whole. This should therefore be regarded as being an exploratory study, and so it will be up to other larger scale studies to test the wider generalizability of the findings reported on here.

5.5 Reporting Mechanism

The overall outcome of the study has been to develop mechanisms to improve and increase the availability of job-oriented courses and skills training, including for green occupations. The ADB goal is to see an increase and improvement in the number of courses and training curricula in skills development, including skills for green occupations, in the four countries by 2017.

5.6 Availability of Detailed Country Case Studies

The full detailed case studies of India, Indonesia, Sri Lanka, and Viet Nam are available for viewing and downloading from, www.cna-qatar.com/research/unesco-unevoc.

Chapter 3
Summary of Main Research Findings: India, Indonesia, Sri Lanka, Viet Nam

Abstract Background and an overview of each of the four case study countries are presented. Details are provided and discussed regarding the main findings for each country regarding environmental policy frameworks, skills development efforts, industry and TVET roles and linkages, sector issues and main policy and practice recommendations, including for education and training. Implications for the business sector, TVET providers, and government are presented and discussed. Recommendations across all four case study countries are presented, discussed, and compared with particular reference to lessons learnt and emerging most effective practices.

Keywords India · Indonesia · Sri Lanka · Viet Nam · Environmental policy framework · Sustainable practices · TVET linkages · Sector issues · Integrated skills development · Promoting green and sustainable practices · Outcome-based incentives · Sustainability as a social issue · Green narrative · Industry awareness · Paradigm shift in skills development system · Outcome-based incentives · Green practices · Sustainability as a social issue · Downstream and upstream supply chains · Social development policies · Construction · Energy · Transport · Services · Manufacturing · Incentives to promote green skills · Green lines of credit · Private and government training systems · Informal sector · Rural sector · Industry and TVET linkages · Regulatory pressures · Direct and indirect incentives · Motivations and practices of foreign aid investment agencies · Generic content

This ADB research study on Education and Skills for inclusive Growth and Green jobs in Asia focused on examining four countries: India, Indonesia, Sri Lanka, and Viet Nam. The purpose of this chapter is to provide the storylines and a summary of the main findings for each country. Full details on each country case study are available at www.cna-qatar.com/research/unesco-unevoc.

R. Maclean et al., *Education and Skills for Inclusive Growth, Green Jobs and the Greening of Economies in Asia*, Technical and Vocational Education and Training: Issues, Concerns and Prospects 27, https://doi.org/10.1007/978-981-10-6559-0_3

1 India

1.1 Background and Overview

The concept of "green jobs," or environmentally sustainable work, has existed in India since prehistoric times. The Vedic literature elaborates on conserving the *panch tatvas* (five elements—earth, water, fire, air, and sky), and many work practices in India, especially in rural areas, have been inspired from conserving the efficient use of the five elements.

However, modern economic activities, including those in agriculture, have been having an adverse impact on the environment and lack sustainable practices in work environments. While there are many shades of green in agricultural practices, these shades get diluted in industrial practices and urban settings. Although industries in India, especially large corporations, are sensitive about climate change issues, this sensitivity varies across sectors. While the renewable energy sector considers the sector itself to be "green" (which is not necessarily true, because many of the work practices may not be green), the manufacturing sector struggles to put the issue high on its agenda because of competitive and regulatory pressures and increasing costs. The services industry is more of a consumer of the other two economic sectors (agriculture and manufacturing) and therefore needs sensitization around the efficient use of energy, water, consumables, and waste management.

Individuals in smaller manufacturing units and in the informal sector are conscious of the importance of sustainable processes and practices, but in the absence of clear standards or a regulatory framework, the enthusiasm of individuals does not get translated into action at the institutional level, thus limiting the scope of green jobs. Nevertheless, the scope of green skills at all job levels is increasing and is being addressed by most industries.

The same holds true for TVET and skills development providers, who, in the absence of any accreditation or well-defined certification systems, are not clear about effective ways to integrate green skills courses into their programs. The highly regulated government TVET and skills development system does not recognize the scope for developing new programs until a regulatory order is established. Instead, they continue to implement traditional curricula and pedagogy even though the demands of industry are changing. The private TVET system, on the other hand, is much more dynamic and responsive to the demands of industry and has the ability and motivation to change programs, courses, and pedagogy to suit the needs of employers, because ultimately its survival depends on the employability of its students.

While the government can push reforms for formal TVET to implement and can create regulatory pressure for industry, the cause of promoting green jobs and green skills can be championed only by industry—mainly large corporations that have the ability to change to sustainable practices and to invest in green technologies. The initiatives undertaken by large businesses can create a ripple effect across smaller enterprises, businesses, and the informal sector, through their value chains, thereby triggering a more widespread demand for specific green skills.

1.2 Summary of Main Findings

India has a Vibrant and a Stable Economy with a Labyrinth of Reform Policies and Programs

India is one of the fastest growing economies in Asia, with a rising share in world GDP. The rate of GDP growth touched 9% in 2008, while the rest of the world was struggling with a financial crisis. India also presents a unique example in managing an enormous population of 1.2 billion people with large geographical, socioeconomic, and political diversity that is governed through a representative democracy consisting of federal, state, and local governments with differing political ideologies and policy emphases.

Stark inequities exist across the various states, economic sectors, demographic groups, and employment types in India. Some states are relatively wealthy and can be easily compared with fast-emerging economies of the world, and some regions are extremely poor and can be compared to Sub-Saharan Africa in Asia. Overall, living conditions are difficult for many individuals and families, with 33% of households with no electricity, 53% with no toilets on the premises, and 42% with no bathing facility.

Across the economic sectors, nearly 54% of the population of India is engaged in agriculture, which contributes less than 17% to GDP. The manufacturing sector employs nearly 21% of the labor force but has a relatively low contribution to GDP of 18%, whereas the services sector, which contributes over 65% to national GDP, employs just 25% of the labor force. Of a total labor force of 475 million, more than half (55%) are illiterate. Ninety-two percent of the labor force are informally employed in the unorganized sector—in agriculture, building and construction, textiles, retail trades, logistics, and transportation—while less than 10% (30 million) are estimated to be in the organized sector. Across demographic profiles, female literacy is 65% compared with 82% for males. Labor force participation of women came down from 36% in 1983 to 25% in 2011. Women earn 62% of men's salaries for equal work. The majority of the labor force work under exploitative and vulnerable conditions, finding employment through informal means such as word-of-mouth references, and contractors. India ranks low (134th) on the Human Development Index (HDI) among 187 countries (UNDP 2011).

Policy measures in India to safeguard vulnerable situations and promote equity are elaborate and widely spread between the national and provincial legislative systems. The Unorganized Worker's Social Security Act, National Floor Level Minimum Wage, National Policy on Child Labour, and Contract Labour (Regulation and Abolition) Act are just some such measures. In addition, a number of employment generation programs, education schemes, and incentive systems aim to provide a just, humane, and equitable society. The Mahatma Gandhi National Rural Employment Guarantee Act, National Rural Livelihood Mission, and National Skill Development Policy are innovative schemes and policies to reform the old systems and to address chronic issues of low employability, poor productivity, and lack of opportunities.

Environmental Policy Framework is Strong; Implementation is Weak

Although India is one of the lowest GHG emitters per capita in the world, with its emissions being only one-quarter of the global average, as good global citizens the government has proactively developed a comprehensive policy, appropriate legislation, a regulatory framework, and programs for the promotion of energy efficiency, renewable energy, nuclear power, fuel switching, and energy pricing reform, to address GHG emissions. The National Action Plan for Climate Change, and its eight missions, outline multipronged, long term, and integrated strategies for achieving important, high-impact goals in energy efficiency, sustainable habitats, water conservation, and forest management. India has a dedicated Ministry for New and Renewable Energy and is one of the top 10 countries of the world in the production of renewable energy through wind, biomass, and solar means.

In addition, the National Urban Transport Policy spells out strategies designed to improve fuel efficiency, the use of cleaner fuels, and new forms of public transport. The Bureau of Energy Efficiency's labeling program for appliances, the solar lamp program, supercritical units in coal-based electricity production, and sustainable tourism certification match global standards in energy usage, energy conservation, and sustainable development. Regulations in the form of the Integrated Energy Policy of 2006 and the Energy Conservation Building Code mandate the green design of new, large commercial buildings, making energy audits compulsory in large energy-consuming units through a dedicated pool of "certified energy managers" who conduct environmental impact assessments, develop environmental management plans, and report annual energy consumption and energy conservation data.

While the policy frameworks attain global standards, this priority has not trickled down to grassroots implementation levels. The absence of implementation mechanisms and a lack of punitive measures encourage defaulters, and poor incentives dissuade enterprises from compliance.

Skills Development Efforts are Diverse and Varied

The TVET landscape has historically been diverse and complicated, with multiple agencies such as the Directorate General of Employment and Training, Ministry of Human Resource Development (MHRD), Ministry of Rural Development, and nearly 17 other ministries engaged in education, vocational training, technical training, and skills development. The launch of the National Skill Development Policy in 2009 led to the formation of the National Skill Development Corporation (NSDC) in 2009, the National Skill Development Agency (NSDA) in 2013, and the Ministry of Skill Development and Entrepreneurship in 2014. In addition, various industries and industry associations, NGOs, and the aid community have been independently working to achieve skills targets. The cumulative capacity of skills development, through all government programs, is about 10 million people per year against a target of 500 million by 2022 (Planning Commission 2011).

While the primary focus of policy makers and training providers is on basic provisioning and on ensuring that large numbers of beneficiaries get enrolled in skills development programs, the ability of certified graduates to do a job productively and remain employable remains questionable in the absence of quality benchmarks and an adequate control and coordinating mechanism. Further, the lack of linkages and pathways among primary, secondary, vocational skills, and higher education providers affects the motivation of the students and pushes vocational education toward a low status.

The absence of productive linkages between the education system and industry further poses problems with regard to retraining, which, in addition to having cost implications, also affects the productivity of industry and trade. Industry representatives during the ADB-EdUHK survey and interviews agreed that there is a large and burgeoning shortage of skilled workers. In the face of higher demand, coupled with poor training and accreditation mechanisms, the private sector has been increasingly forced to establish its own training and quality control mechanisms.

1.3 Summary of Key Findings from the ADB-EdUHK Survey

Strong Regulation Both Deter as Well as Motivate Sustainable Practices

The ADB-EdUHK survey and interviews refer to a number of findings that have implications for approaches used to address problems in education and training and their alignment with the issue of sustainable growth. For example, strict regulation does not necessarily result in compliance. Rising regulatory requirements have an impact on all industry sectors, and nearly half the companies surveyed believed that being strongly regulated through government interventions was both a deterrent as well as an opportunity to instill environmentally sustainable practices.

While at one level policy challenges exist, such as integrating economic, industrial, environmental, and skills development policies, at another level there are institutional bottlenecks to balancing multiple efforts, ranging from compliance-related matters to ensuring productivity enhancement, to saving costs, to managing innovations to be competitive, and to developing human capital.

Rising Costs and Short-term Targets Push Environmental Management Low on the Priority List

Short-term targets often overshadow long-term outcomes in both the government and the private sectors. Rising costs, increased competition, and poor profitability outweigh the efforts of technological innovations, skills development, and sustainable practices. While an institutional approach to sustainability is yet to evolve, individual motivation is high. Close to 80% of industry respondents witnessed a growing awareness among senior managers concerning climate change or environment-related jobs, while only 50% agreed to integrate jobs related to dealing with climate change into the work environment.

Shortage of Skilled Workers is Impacting Businesses

Nearly 60% of respondents from the energy sector confirmed that they are facing a shortage of skilled workers in energy-efficient construction and retrofitting, renewable energy, energy efficiency, and environmental services, thereby impeding the transition toward green growth. Rural electrification programs are suffering because of a lack of skilled workers. The reasons for these shortages include a scarcity of scientists and engineers, the poor reputation and limited attractiveness of certain sectors such as waste management, and the limited number of well-qualified and high-quality teachers and trainers in environmental services.

Corporate and Social Responsibility Mandates and Import-Export Requirements are Key Drivers to Develop Green Jobs and Green Skills

The hospitality industry appears to be most active in adopting sustainable practices in its businesses, whereas the transport and logistics sector leads in employing new workers for climate change. Corporate and social responsibility (CSR), import-export requirements, and government legislation seem to be the most important drivers for developing new green jobs and skills. While the demands for green skills in traditional industry arise mostly out of compliance with regulations, demands in new industries arise out of the resource crunch and global sustainability arrangements such as renewable energy. While the scope of green jobs may be limited, the scope of green skills in all jobs is unlimited. There is a greater emphasis on the requirements of hybrid skills (generic plus green) that top-up existing skills sets.

Industry—TVET Linkages are Weak and Cosmetic

The low levels of engagement between industry and TVET institutions continue to deter the improvement of courses and development of industry-relevant curricula. On average, only 18% of industry respondents provide course feedback to the TVET institutes, while TVET providers appear to be unsure about their programs with respect to green skills content. Green topics are largely excluded from the skills development curriculum. Only 50% of TVET providers have "shades of green" content in their general curricula. Large numbers of nongovernment and private institutes seem to have created their own internal certification criteria, which may not be necessarily recognized by national and authorized bodies such as the National Council for Vocational Training or the All India Council of Technical Education. Most institutions in the survey indicated that their teachers are not adequately trained to provide green skills training to students. The absence of accreditation and certification on green standards is a deterrent.

Sector Issues

Construction. Energy management and sustainable construction continue to be important concerns of the sector. Time and cost overruns are one of the major problems of the industry, triggered by a lack of formal training and inadequate systems related to skills assessment and the certification of construction workers. The construction sector is expected to employ about 92 million people by 2022 from the current 33 million.

Nearly 40% of respondents in the ADB-EdUHK survey agreed that they needed new workers for jobs and employment related to climate change; nearly 50% agreed to integrate jobs related to climate change in the work environment; and nearly 80% agreed that there is a growing awareness among senior managers about issues related to climate change. Nearly all respondents rated innovation in technology, products, processes, and services as important to be able to manage waste at construction sites and to improve energy efficiency. The change in work environments is leading to demands in generic and specific green skills.

Generic green skills (a foundation for all workers) required are awareness about environmental issues, energy and water efficiency, waste reduction and waste management, green procurement, understanding of standards and legislation, and the ability to implement them. Specific green skills required are in installing pre-fabricated components for green buildings, technical skills to promote heat and energy efficiency, energy-saving practices, and safety for firemen in brick kilns.

Energy. While the consumption side of the energy sector such as the labeling of equipment and appliances, energy efficiency in buildings, energy efficiency in industry, residential lighting, and agricultural pumping have identified roles and jobs that require green skills, the energy production side (largely coal-based) is yet to move to efficient technologies that can reduce emissions, and so the requirements of green jobs and skills are yet to be articulated. The renewable energy sector, on the other hand, does not identify green skills as an "add-on" in the industry, because they regard the sector itself as green and consider all jobs as green jobs. However, other problems are being faced by the renewable energy sector.

A Confederation of Indian Industry (CII) report (2010) identified that salary packages in the renewable energy industries are less when compared with other industries, and therefore attracting skilled labor is getting increasingly difficult. This was also observed in the ADB-EdUHK industry interviews, where graduates with renewable energy knowledge preferred IT jobs over positions in renewable energy because of the remuneration difference and relocation to less attractive locations.

While energy sector policies and strategic direction have sparks of excellence, the corresponding requirement for green skills is on the rise. Skills in project management, installation, commissioning, and grid integration of large-scale renewable energy projects, wind resource assessment, and technocommercial marketing skills are some of the skills increasingly required in wind energy. Some of the solar energy sector skills required are installing building integrated photovoltaics (BIPV) in buildings; project development skills in handling concentrated solar collectors (CSP); system integration in solar photovoltaics (PV); construction, installation, and commissioning of solar thermal systems; operation and maintenance skills in trouble shooting of circuitry of solar PV lanterns; and home lighting systems.

Apart from renewable energy, coal-based power generation requires skills in operating and managing supercritical units, ultra supercritical units, and underground coal gasification. Skills in carbon capture and storage, expansion of non-conventional gas resources like shale gas and coal bed methane, and coal washing are also required.

Transport. While Indian Railways is one of the largest employers in the world, having a separate ministry, roads continue to be the dominant mode of transport in India, carrying 85% of passenger and 60% of freight traffic, imposing a high cost on the economy by way of a much higher dependence on fossil fuels and high levels of GHG emissions.

While vehicular emissions standards get progressively tightened, compressed natural gas (CNG) is rapidly being introduced for public transportation. Allied to the transport sector is the automobile and automotive component sector, which provides direct and indirect employment to more than 13 million people. Due to this rapid expansion and change in technologies, the transport sector, which currently employs nearly 7.3 million persons (more than 90% in roads and railways), may need 17.7 million employees by 2022 for roads, ports, railways, airports, and warehousing (NSDC 2009). In addition, the automotive industry may, directly and indirectly, employ 35 million by 2022.

In addition to quantitative expansion, the industry is also experiencing qualitative changes. Nearly 80% of ADB-EdUHK survey respondents agreed that there have been job role changes including knowledge skills, technical skills, and attitude. The changes are demanding skills in different forms for lower as well as higher level operations. For example, rapid mass transport needs top-up skills in driving safety, communications, using complex machinery, and handling hazardous materials. Freight transport includes skills in materials handling equipment such as stackers, hydraulic or hand pallet trucks, forklift trucks, and jib cranes; handling of hazardous materials; and ability to handle increased tonnage and higher capacity trucks.

Services. Hospitality (travel and tourism) remains one of the largest service industries in India contributing 6% to national GDP and 9% to total employment. It is one of the priority sectors for the government and industry. The total number of jobs (direct and indirect) in the tourism sector in 2016 is estimated to be 77.5 million, up from 53 million in 2010.

The issue of environmental sustainability is actively being pursued by large and organized corporations; however, the same is yet to be percolated to small and medium-sized enterprises and the informal sector. While a number of standards and certifications such as Green Globe, ISO 14001, Earth Check, and Sustainable Tourism Eco-Certification are available for larger corporations, these are yet to be adopted by smaller enterprises.

Even though an increasing number of enterprises are adopting energy-efficient technologies, there are no specific requirements for specialized people for building a green environment in companies. The enterprises largely depend on top-up skills for existing staff. This was also seen in the ADB-EdUHK survey, where less than 50% of respondents expressed the view that a demand for green skills has impacted their business. The concern for increasing costs, changing consumer demand, and government legislation was much higher.

Some of the top-up green skills that the sector is actively seeking are recycling, reusing, and reducing fuel consumption in operations; disposing of non-biodegradable garbage in a responsible way; procuring green materials; converting the use of polythene bags to paper bags and cloth bags; and using biodegradable cleaning chemicals. Some other emerging roles and green jobs in the sector are home stay owners and/or workers, golf caddies, polo grooms, nature guides, rural and water and hill adventure guides, security guards, tourist vehicle drivers, and masons for heritage constructions.

1.4 Recommendations

While the demand for green skills in all jobs is increasingly expressed by industry, not much interest is shown in differentiating between green and nongreen jobs. Though academic efforts may continue to identify green jobs across sectors, the industry sectors are more interested in introducing sustainable practices in production and services, which may have implications for creating new job roles that are green, or the requirement of top-up skills in the existing workforce. The evidence is clear that skills shortages are impeding the profitability, productivity, and competitiveness of enterprises.

Green skills development can happen at formal TVET institutions, through in-house skills upgrading, or as a response large societal movements, such as those lobbying about environmental improvement and sustainable development issues. However, each of these methods has limitations related to relevance, credibility, and cost-benefit that policy makers, industry leaders, and civil society organizations have to address in a coordinated way. For example, TVET cannot work in isolation, since it has to align with economic policies, climate change, and industry requirements.

Integrate Skills Development Policies with Industrial, Economic, and Environment Policies

While the National Skill Development Policy of 2009 is relatively new, the industrial policies and environmental policies have a legacy of their own and have evolved according to the socioeconomic and political realities of their formulation. Institutions have evolved around these policies over the years, and it is not an easy task to build consensus across policies and institutions to have a common approach with respect to green and inclusive growth.

An immediate assessment is required of the elements such as research skills in solar energy production under the National Solar Mission, and technical skills in recycling materials and urban waste management under the National Mission on Sustainable Habitat to determine the extent and nature of possible policy integration. The assessment may study important policies, programs, and institutions; explore areas of convergence and coordination; and provide the necessary thrust for green jobs and green skills development. Government, with the help of multilateral agencies, can bring in international expertise and good practices in order to learn the best ways and methods of policy rationalization and capacity development of institutions and individuals. Capacity development on processes and tools for green skills of institutions such as NSDA, NSDC, Aajeevika skills, the Capacity Building for Industrial Pollution Management Project, and the National Green Tribunal can also be organized under the leadership of the respective ministries.

The key responsibility lies with the Ministry of Skill Development and Entrepreneurship along with the Ministry of Environment and Forests and the Ministry of Commerce and Industry.

Promote Green and Sustainable Practices as a Business Issue

The pressure of achieving business targets, and stiff competition, push the issues of sustainability lower on the priority list, especially in small and medium-sized enterprises (SMEs) and in the informal sector. Businesses fail to realize that sustainable practices can solve problems of increasing costs, provide an edge over competition, and enable compliance with climate change mandates. Greening, therefore, should become a business issue and be championed by larger companies.

Large corporations such as Reliance Industries Limited and the Tata Group can help champion developing policy on green standards and sustainable practices. Green practice standards can also be included in the industrial policies of new economic corridors.

Further, they can communicate widely and strongly the benefits of adopting sustainable practices, and upskilling the existing workforce in SMEs through existing supply chains. Models of green innovation, as well as green skills development, may be rewarded and recognized through initiatives supported by CSR funds and also may be patented.

In addition, the larger corporations can also partner with leading higher education institutions in India as well as abroad through well-funded research projects on green technologies and practices that may need specific green skills.

The key responsibility rests with large corporations such as the Tata Group or Reliance Industries or with public sector corporations such as Indian Oil Corporation Ltd., Bharat Heavy Electricals Ltd., and Oil and Natural Gas Corporation along with CII and the Federation of Indian Chambers of Commerce and Industry (FICCI).

Coordinate Action Between Foreign Aid and Investment Agencies

The ILO Decent Work Team of South Asia can closely work with the World Bank's Vocational Training Improvement Project to bring elements of sustainable skills to TVET institutions. Similarly, the Department for International Development of the United Kingdom, UNDP, Japan International Cooperation

Agency, GTZ, Australian Agency for International Development, and ADB can come together to develop a consolidated national strategy for support toward green growth, green jobs, and green skills.

Existing ADB support for strengthening skills development efforts in the states of Meghalaya, Kerala, and Odisha can have elements of green skills development. ADB technical assistance to NSDA can help focus the National Skills Qualifications Framework (NSQF) to build in components of green standards. ADB technical assistance to NSDC can help sector skill councils to develop green skills standards.

Documentation and dissemination of good practices (national and international) can be supported by aid agencies, while multinational companies (MNCs) may support models of innovation toward green practices and green education with private sector agencies and TVET institutions.

The key responsibility lies with aid agencies: ADB, World Bank, UN, MNCs through the CII, and FICCI.

Initiate Easily Implementable Activities Around Green Skills

Generic skills in training are easy to introduce and can have a progressive path to specific skills and courses. Content related to sustainable practices and climate change issues can be introduced into existing curricula, not only in TVET institutions but also through school education and higher education institutions. Major government-funded training schemes such as Aajeevika Skills (National Rural Livelihoods Mission) and the STAR scheme (NSDC) can include a number of "green" occupations in their training priorities.

Higher education institutions may also take the lead in the professional development of trainers through in-service training using special modules on green skills. The teacher training programs and courses can also start including generic and specific green skills modules. These can be included as accreditation requirements for training programs. MHRD may establish a green benchmark for teacher training programs.

The key responsibility is with the Ministry of Skill Development and Entrepreneurship (MOSDE) along with NSDA, NSDC, CII, and FICCI.

Increase Coordination and Rationalization of the Private and Government Training System

Training by the private sector cannot be ignored. While reforms in government TVET are being led by the line ministries, reforms in the private sector can be pushed through industry and employers. Parity in quality of training, incentives, governance systems, and industry relevance is required between the two. In addition, the state should create a framework to accredit and recognize the different forms of vocational training undertaken by different private industries.

While institutional initiatives are quicker to implement, large-scale reforms may also be initiated simultaneously. For example, MOSDE can develop green skills standards and dovetail them with the NSQF and National Occupational Standards (NOS). The NSDC may add "green skills" criteria in the validation process for NOS.

Training providers can also be colocated with the industry to improve the relevance of training, improve teacher industry experience, offer research and development opportunities, create more responsive delivery arrangements, and foster industry skills development clusters.

The key responsibility is with MOSDE along with NSDA, NSDC, and large private sector corporations. Also to be engaged are the Ministry of Industry and Commerce; Ministry of Corporate Affairs; Ministry of Heavy Industry and Public Enterprises; and Ministry of Micro, Small and Medium Enterprises.

Introduce Outcome-Based Incentives for SMEs and TVET Institutions

In a scenario where there is great competition, it is unreasonable to expect that SMEs would be self-motivated to promote sustainable practices. India has an estimated 1 million such enterprises, and so reaching out to such a large number of organizations can be possible only from outcome-based incentives that could be direct, such as tax breaks; subsidies; viability gap funding; or indirect benefits such as awards and recognition, land provision, provision of training equipment, and international exposure visits. In addition, enterprises offering green skills development training should be entitled to general incentives concerning vocational training under government regulations or be allowed to recover training expenses through users or sponsoring institutions.

In parallel, TVET institutions that initiate programs or make special efforts for green skills development could be motivated through special entitlements, additional budgetary support, technical assistance, and/or international industrial exposure.

The key responsibility lies with the Ministry of Finance, along with the Ministry of Commerce and Industry, MOSDE, and Ministry of Labour and Employment.

Promote Green Practices and Sensitivity Toward Sustainability as a Social Issue

Green skills development should be understood not only as a matter of curriculum, and, in terms of content, limited to TVET institutions, but as something that should be built into a larger ecosystem of education, workplaces, social interactions, and political debates. In addition to formal TVET, sensitivity toward the environment should be developed through social interactions and public communications.

All forms of media—whether print, radio, television, or social—can be encouraged to position intelligent messages so that the community at large sees value in the importance of society moving toward greener practices, with green skills development becoming a necessity rather than an option. NGOs and self-help groups (SHGs) can be targeted with special green skills modules to reach out to involve women and marginalized groups, and to encourage them to develop green skills, which may or may not be employment-related skills.

The key responsibility will be with MOSDE along with CII, FICCI, the News Broadcasters Association, and the India Broadcasting Foundation.

Expand the Scope of Green Jobs and Green Skills to the Informal and Rural Sectors

While most engagement between industry and the government with respect to sustainable growth is limited to the formal and organized sector and addresses mainly urban issues, research and policy must also generate evidence on the need for and situation of environmentally sustainable jobs and skills with respect to work in the informal economy such as small-scale manufacturing, trading, art and craftsmanship, and related to work in rural areas, especially agricultural processing and other nonfarm-related work.

The key responsibility lies with multilateral agencies such as ADB, the World Bank, and ILO along with research institutions and universities.

2 Indonesia

2.1 Background and Overview

In Indonesia the concept of greening is interpreted differently in the realm of TVET. While for some providers, greening is viewed as being a part of hygiene and safety, for others it relates to the introduction of a new curriculum relevant to international environmental standards, waste management, and pollution prevention, or else it has a broader interpretation that relates to climate change issues. The strongest point of agreement across all TVET providers concerns the need to develop a comprehensive approach toward adjusting TVET to meet the needs of a greening economy and the need to offer high-quality and relevant initial and continuing training programs to meet the demand for green skills. The fragmented understanding of the concept of greening is due mainly to the absence of a regulatory framework for green skills development in the country.

Industry is grappling with economic pressures and feels unable to address matters related to sustainable production; thus the need for specific activities concerning green skills is relatively low in industry's order of priority. However, there appears to be a contradiction, since at the same time industry has unequivocally stressed the need for generic green in all types of jobs. Generic green skills can be topped up, over the standard skills, and can be imparted either through TVET institutions (in the case of existing students) or through in-service training for members of the workforce through external experts or short-term development programs in TVET institutions. Such an approach could also be a revenue-generating opportunity for the otherwise financially ailing TVET institutions.

TVET and the industries have created impressive "islands of excellence" that need to be mainstreamed. At the same time, the recently elected new government is currently positioning itself to prioritize its efforts toward giving a new direction to green growth with more integrated policies, leveraging upon the international expertise and the progressive work of past governments. Efforts in the past such as the "factory in schools" or "Esemka cars[1]" initiatives are to be popularized, and schools adopting such practices need to be incentivized. Similarly, Indonesia has a unique system where higher education is closely integrated with the vocational system, and such a link needs to be taken advantage of to develop high-quality trainers for the TVET system and to promote high-quality research in green practices.

Although green jobs are an emerging subject in Indonesia, the number of initiatives by the government, multilaterals, the private sector, and civil society in green jobs and green skills development are like scattered pearls that need to be beaded together. There are different "shades of green" in each of these initiatives, and some are very specific to a particular industry, geographical area, or situation; however, the real challenge is to bring out the principles behind these successes and to replicate them on a larger scale across industries and geographical regions. Large industrial corporations, together with international community support, can champion such efforts to make green practices a habit for individuals, businesses, TVET society, and society at large.

2.2 Summary of Main Findings

Indonesia is slowly recovering from a recent economic setback in 2013 when the local currency (the rupiah) suddenly fell nearly 20% against the US dollar. The GDP growth rate slowed from a peak of 7–5%; and there was a spike in inflation, which rose to more than 8.5%, from an earlier level of 4%. Economists argue that the current economic crisis in the country may just be short term and will be corrected if restrictive policies on matters such as mining exports and foreign investment rules are revisited and made more progressive.

With a population of nearly 250 million (the fourth largest in the world), Indonesia achieved considerable economic progress after attaining democracy in 1999, through the progressive policies of the government and the optimistic mood of the people. Policies such as the Master Plan for Acceleration and Expansion of Indonesia Economic Development, which were implemented with a spirit of "not doing business as usual," not only contributed to economic progress but also had a positive impact on social indicators. Since then per capita GDP has increased fourfold, poverty has halved, literacy rates have reached an all-time high (95.8%), and the HDI improved to 0.62 from 0.54 (a 15% rise) in just 10 years.

[1]Esemka cars (PT Solo Manufaktur Kreasi—SMK cars) are cars locally developed by senior high school students from several SMK schools (Sambijantoro 2015).

Eighty-two percent of Indonesia's economic activity is concentrated in the western region of the country (Bali, Java, and Sumatra). Seventy percent of the population lives on the island of Java, which accounts for a mere 6% of the land area of the country. In recent years, several million migrants from other islands have moved to Jakarta and other major cities in search of employment but have found jobs mainly in low-end services, becoming part of a vast informal economy, which accounts for close to 70% of the GDP. They rarely earn the official minimum wage and receive few government benefits.

Existing education levels of the nearly 118 million labor force (38% of whom are women) are worrying, with only 31% having attained secondary education. Almost half of the labor force (48%) have not successfully completed primary education. Low levels of education, coupled with the informal nature of many areas of employment, result in a large portion of the workforce being vulnerable to exploitation and poverty.

2.3 *The Green Narrative—Dispersed and Disconnected*

The downside of the economic turnaround in Indonesia since achieving independence has been its adverse impact on the environment. Carbon dioxide (CO_2) emissions grew three-fold between 1999 and 2012. Furthermore, natural disasters have subjected the country's flora, fauna, and mineral resources to significant risks. Subsequently, the country has signed multiple international environmental protection agreements to protect biodiversity, lessen desertification, reduce ship and sea pollution, effect hazardous waste management, and promote marine life conservation. The National Council on Climate Change (NCCC), chaired by the president, is the main government body responsible for climate change policy coordination and development.

Several presidential regulations and laws, ranging from laws on the regulation of CO_2 emissions, waste management, renewable energy, ecotourism, and community training on environmental sustainability in exploration, to the exploitation and rehabilitation of the environment have been passed, although large-scale implementation has yet to occur.

The concepts of green workplaces, green jobs, and green skills are somewhat new in Indonesia, and there is no proven set of methods, policies, or approaches to successfully implement green initiatives. However, organizations such as ILO have been providing technical assistance concerning green jobs in the tourism sector with active support from the Ministry of Manpower and Transmigration (MoMT), the Ministry of Tourism, and NCCC.

The private sector is also contributing to efforts toward green initiatives and practices. The Green Building Council of Indonesia has created GREENSHIP certification; Indecon (Indonesia Ecotourism Network) promotes and develops ecotourism in Indonesia; and garment factories are closely regulated by international consumer mandates.

2.4 Education and Training: Awaiting Deeper Reforms but Making Efforts Toward Green Skills

Unemployment among those who have attained secondary education (vocational) is nearly 12% compared with the country average of about 6.25%. The government's policy push, and a budget allocation of almost 20% of GDP for education have not only helped expand the network of educational institutions across 6000 inhabited islands but have also improved the progression of cohorts. For example, those achieving lower secondary education went up from 31% in 1994 to 52% in 2013.

TVET is spread across three levels: secondary, postsecondary, and tertiary education. TVET depends more on private (nearly 73%) rather than government institutions. The 16 university-level institutions that offer TVET teacher education plus the inclusion of vocational courses at the undergraduate and graduate levels have created a structure that is unique to the Indonesian education system.

Generally TVET is not as attractive an option as an academic pathway; however, expanding TVET at the higher education levels and the setting up of the NSQF are important breakthroughs in enhancing the image and attractiveness of TVET in Indonesia. The challenges of image concerning TVET in society and industrial communities have been improved over recent years due to committed intervention by the government, resulting in TVET being considered as a viable option for youth as an educational pathway.

The education system as a whole is making efforts to incorporate green content into training programs. Various organizations and ministries have piloted projects and initiatives to raise awareness and mainstream the green agenda. For example, senior secondary schools encourage students to green their campuses. The Ministry of Youth and Sports, together with the Central Bank of Indonesia, supports the "Indonesian green entrepreneurship program," which is aimed at increasing levels of entrepreneurship and the greening of businesses. The NCCC supports capacity building through a network of 19 universities, through courses aimed at lowering pollutant emissions. Similar pilot projects have been initiated by the Ministry of Education and Culture (MoEC) and the MoMT. The 2013 New Curriculum includes some green skills in the health and safety modules. For example, refrigerant recovery from air conditioners, and ways to process electronic waste are included as part of safety competencies; and energy saving and waste recycling are addressed in many major subjects.

2.5 *Summary of Key Findings from the ADB-EdUHK Survey*

Industry Awareness About Green Skills is High for Internationally Linked Companies

While the concept of green skills for all jobs is more widely understood than the concept for green jobs, companies that have strong international links tend to be the ones that most appreciate and want to implement sustainable practices. These companies also possess some form of international and national environmental accreditation, value customer demands and sensitivity to green issues, and encourage compliance across their supply chains. More than 50% of respondents in the construction sector, and over 40% of hospitality sector respondents agreed that a lack of skilled employees is proving to be detrimental to their businesses. However, fewer than 30% of respondents in all sectors agreed that they are currently employing new workers (or training workers) for green jobs.

SMEs Have Less Motivation Toward Sustainable Practices

Fewer than 30% of businesses across sectors, according to the ADB-EdUHK survey, use sustainable practices such as energy-efficient products and services, and recycling, in their work practices. Government legislation, increasing costs, and rising industry standards are more pressing issues affecting the businesses surveyed. However, more than 40% of the businesses surveyed in construction, hospitality, manufacturing, and energy expect high standards of sustainable processes and infrastructure across their supply chains.

Engagement Between Industry and TVET Institutions is Low

While a large number of industry respondents in the ADB-EdUHK survey agree on the importance of TVET work-based training of potential employees, the level of engagement between businesses and TVET institutions is reported to be low. Fewer than 40% of respondents expressed satisfaction with the outputs from TVET institutions.

TVET Institutes Face Funding Constraints to Implement Green Skilling

According to the ADB-EdUHK survey, more than 80% of vocational secondary schools (VSSs) and higher education vocational training providers (HE-VTPs) agreed that a transition to the greater use of products and services that comply with environmental regulations and standards is impacting changes in their programs and courses. Eighty-two percent of VSSs and 74% of HE-VTPs recognize energy efficiency as a driver toward changing programs and courses. However, lack of funds is a constraint in implementing green training. A small number of partnerships with the private sector initially provided solutions, but these could not be sustained due to a lack of recurring funding. For example, Toyota installed oil recycling equipment at one VSS and conducted teacher training programs, but over time the equipment could not be maintained due to lack of funds, pushing the VSS back into using more traditional methods of teaching.

Sector Issues

Construction. According to a McKinsey report (see Mourshed et al. 2012), the construction sector in Indonesia will find it difficult to fill 40–50% of positions requiring tertiary education and at the same time, it predicts an oversupply of TVET graduates working in the sector. More than 50% of respondents could not meet their sustainability goals due to lack of skilled workers in environment-related jobs. Growth in the construction sector in recent times, together with a rise in awareness about environmental processes and practices, has contributed to job changes, especially with regard to the knowledge, skills, and attitudes of workers. Nearly 73% of businesses in the construction sector identify the introduction of new standards related to sustainability or environmental issues as leading to new skills.

Green skills are increasingly required in impact assessments, innovative construction materials, and building techniques, and with regard to training in the correct use and application of materials and retrofitting. Some other areas of skills requirement are green certification, green construction techniques, and waste management at construction sites.

Energy. Indonesia is aiming to increase its renewable energy base to 26% by 2025 from the current level of 16%. There is enormous pressure building on existing enterprises to increase their capacity and move toward greener practices. Nearly 75% of respondents in the ADB-EdUHK survey agreed that workloads were higher than usual and companies were not working to their full potential. In addition, government legislation such as PROPER is also creating pressures to incorporate new skills into work environments.

Skills are required not only at the operational level but also at higher levels, such as skills to assess business performance against the PROPER criteria, laboratory skills to monitor water and air quality, water and waste management, and stakeholder management. The demand for operational skills relates mostly to using vendor equipment, waste management, and top-up skills in installing energy-saving and alternative energy devices.

Transport. Transport contributes to nearly 13% of GHG emissions in Indonesia. In addition, nearly 51% of oil consumption in the country is attributed to the transport sector, while the number of vehicles is predicted to double by 2035. Road transport accounts for approximately 70% of freight movement and 82% of passenger movement. As a result, air quality has degraded to the extent that 60–80% of air pollutants in metropolitan cities are attributed to vehicular emissions.

While the government is taking active policy steps to address these problems, such as tax-exemption initiatives to encourage the production of low-cost green cars, and by mandating stricter emission norms, the sector needs urgent reforms that are implementable and also needs a sensitive and appropriately trained workforce to counter growing environmental concerns arising out of the sector. Some specific skills required in the sector include top-up skills in driving safety, communications, using complex machinery, and handling hazardous materials. Skills are required in materials handling equipment such as stackers, hydraulic or hand pallet trucks, forklift trucks, and jib cranes; dealing with hazardous materials; and an ability to accommodate increased tonnage and higher capacity trucks in freight transportation. Skills required in higher level jobs include understanding of fuel efficiency technology, carbon emissions, and sustainable urban transport planning.

Services. Indonesia is an attractive destination for tourists, with the hospitality industry growing at a rapid pace. The number of jobs in the sector was expected to reach 9 million by 2015, and the government plans to allocate $140 billion for infrastructure development to support the projected expansion of the hospitality industry.

With nearly 60% of survey respondents agreeing that jobs in the industry will change due to a rise in sustainable practices, and more than 90% agreeing with the importance of environmental awareness in the sector, there is a corresponding requirement for skills in the areas of waste management, energy-efficient buildings and appliances, and the use of water. Some top-up green skills that the sector is actively seeking are recycling, procuring green materials, converting the use of polythene bags to paper and cloth bags, and using biodegradable cleaning chemicals.

Manufacturing. Manufacturing accounts for 25% of Indonesia's GDP. According to a World Bank (2012) report, 84% of employers in manufacturing have difficulty filling management positions, and 69% have problems sourcing other skilled workers. This is a serious issue that inhibits the further development and expansion of the sector. In addition, regulatory requirements and competitive and societal pressures are directing manufacturing industries to adopt greener and environmentally sustainable technologies and practices. Nearly 76% of respondents in the ADB-EdUHK survey identified job changes as being the most significant change related to greening their businesses, and 78% agreed with the importance of TVET work-based training of potential employees.

Training is required in waste disposal and recycling, including knowledge of government regulations and related compliance. Skills gaps are reported in handling hazardous materials, LEAN manufacturing, and undertaking environmental impact assessments.

2.6 *Recommendations*

While the training providers integrate areas of green skills into their courses and curricula, the actual green practices required in agriculture, industry, and the services sector are quite different. The agriculture sector has connotations of "green" such as water conservation, the use of fertilizers, and appropriate cropping patterns, while the services sector is concerned more with conserving energy. The industry sector is interested in processes such as pollution control, recycling, waste management, procurement, and energy audits. The demand for green skills in traditional industries arises mostly out of the need to comply with regulations. Green skills demands in new industries such as renewable energy production arise out of a need to conserve resources and due to global sustainability agreements. Within the industry sector, there are different "shades of green" among the subsectors that need to be considered when developing strategies.

Introduce a Paradigm Shift in the Skills Development System to Align with Economic and Environmental Policies

While climate change commitments, laws on environmental protection, and compliance with emission standards take shape in Indonesia, the education and skills development system need a paradigm shift in order to align with the growing needs of industry for environmentally sustainable skills in every job role.

A closer assessment is needed to understand the elements of skills development policies and the scope for their alignment with the industrial, economic, and environmental policies of the country. Such an assessment may study important policies, programs, and institutions; explore areas of convergence and coordination; and provide the necessary thrust for green jobs and green skills development. The government, with the help of multilateral agencies, can bring in international expertise and good practices in order to learn the best ways and methods toward policy rationalization and the capacity development of both institutions and individuals.

The key responsibility rests with the National Board on Development Planning (BAPPENAS) along with MoEC.

Design and Provide Outcome-Based Incentives for the Industry (Mainly SMEs and TVET Institutions)

In a scenario of fierce competition and thin profit margins, it is not realistic to expect that industries, especially SMEs, will be self-motivated to promote sustainable practices. While the large industrial corporations have the ability to invest in sustainable practices and technology, the same is not true for smaller enterprises. Outcome-based incentives are therefore required to motivate SMEs to adopt greener practices. Incentives could be in the form of tax breaks; subsidies; viability gap funding; or indirect benefits such as awards and public recognition, land provision, provision of training equipment, and international exposure visits. In addition, enterprises offering green skills development training may be entitled to general incentives concerning vocational training under governmental regulations, or be allowed to recover training expenses through users or from sponsoring institutions.

In parallel, TVET institutions that initiate programs or make special efforts for green skills development could be motivated through special entitlements, additional budgetary support, technical assistance, and/or international industrial exposure.

The key responsibility is with the Ministry of Finance, along with the Ministry of Industry and Trade and MoEC.

Promote Sustainable Practices as a Business Issue

The ADB-EdUHK study clearly shows that most businesses fail to realize that sustainable practices can solve the problems of increasing costs, developing a positive edge over competitors, and compliance with climate change mandates. Efforts to become green should become a business issue and move up to become higher on the priority list for action, even though the pressures to achieve business targets and to successfully meet stiff competition are high, especially in SMEs and in the informal sector. Larger companies can prove to be role models and become champions in advocating the benefits of adopting sustainable practices, and ups-killing the existing workforce, through communication and regulation across existing supply chains.

The key responsibility lies with large corporations such as Astra International and the Lippo Group or public sector corporations such as Perusahaan Gas Negara, Perusahaan Listrik Negara, and Pertamina, along with the Indonesia Chamber of Commerce (KADIN).

Push Green Growth, Green Jobs, and Green Skills Through Foreign Aid Agencies' Coordinated Action

ILO is working with local communities on the Indonesian part of Kalimantan on a green jobs program called the "Green Livelihood Access for Central Kalimantan's Inclusive Environmental Response to Climate Change" that can access UNDP funds for energy conservation and environmental initiatives. There is the GIZ PAKLIM program to implement and disseminate climate mitigation and adaptation measures. Conservation International is working in the Sundaland areas, Wallacea areas, and the tropical wilderness areas of Irian Jaya in the hospitality and tourism sectors and can build synergy with the other international initiatives of ADB and the World Bank in the country.

Documentation and dissemination of good practices (national and international) can be supported by aid agencies, while MNCs may support models of innovation toward green practices and green education with private sector agencies and TVET institutions, respectively.

The key responsibility is with the aid agencies—ADB, ILO, GIZ, World Bank—along with KADIN and BAPPENAS.

Introduce Generic Skills with Progressive Paths to Specific Skills into the Curriculum

Generic skills in the standard curriculum of training are easy to introduce and may lead progressively to specific skills and certification. The introduction of such skills may be targeted through the full cohort of school education, technical education, and higher (university) education. Simultaneously, research in higher education institutions may be encouraged toward newer industry practices and technologies that are environmentally sustainable. Higher education institutions may also take the lead in the professional development of trainers through in-service training.

While institutional-level initiatives are quicker to implement, large-scale reforms may also be initiated in the form of aligning the New Curriculum of 2013 (Government Regulation No. 70, 2013) and National Education Standards with the NSQF. Ways to include green skills standards in the Indonesian National Standards also need to be explored.

The key responsibility is with MoEC.

Align and Rationalize the Standards and Outcomes between Large Private Training and the Government Systems

Vocational training by large private sector institutions in Indonesia is very important and cannot be ignored. While reforms in the government TVET system are being led by the line ministries, reforms in the private sector can be pushed through by industry and employers. Parity in quality of training, incentives, governance systems, and industry relevance is required between the two. In addition, the state should create a framework to accredit and recognize the different forms of vocational training undertaken by private providers and industry.

The key responsibility lies with MOEC together with large private sector corporations.

Promote Green Practices and Sensitivity Toward Sustainability as a Social Issue

Green skills development should be understood not only as a matter that concerns the curriculum, and in terms of content limited to TVET institutions, but also as something that should be built into the larger ecosystem of education, workplaces, social interactions, and political debates. In addition to formal TVET, sensitivity toward the environment should be developed through fruitful social interaction and public communications.

All forms of media—print, radio, TV, and social—can be encouraged to position intelligent messages so that the community at large sees value in the importance of society moving toward greener practices, with green skills development becoming a necessity rather than an option. NGOs and SHGs can be targeted with special green skills modules to reach out to women and marginalized groups, and to encourage them to develop green skills.

The key responsibility is with MoEC along with KADIN and large media companies.

Expand the Scope of Green Jobs and Green Skills to the Informal and Rural Sectors

While most engagement with industry and government with respect to sustainable growth is limited to the formal and organized sector, and addresses mainly urban issues, research and policy also need to generate evidence on the needs and situation of environmentally sustainable jobs and skills with respect to work in the informal economy such as small-scale manufacturing, trading, and art and craftsmanship, related to work in rural areas, especially concerning agricultural processing and other nonfarm-related work.

The key responsibility is with multilateral agencies such as ADB, World Bank, and ILO along with research institutions and universities.

3 Sri Lanka

3.1 Background and Overview

At the individual level in Sri Lanka the concept of "green jobs" and related "green skills" is understood mainly in terms of the importance of saving natural resources and not polluting the environment. In large industrial corporations, sensitivity toward "greening" is generally high because of a concern about rising costs. In the case of individuals in small, medium, and micro enterprises, although motivated, they do not generally find a platform, or an institutional mechanism, to translate their enthusiasm and understanding of sustainability into practice.

The importance of practices in support of sustainability in the government system is triggered by sensitivity toward the environment and global climate change agreements. Policies toward sustainable development are having a gradual impact on education and training. For example, TVET institutions have started including specific teaching and learning modules with respect to sustainability in their education programs, although in the absence of suitable standards and certification, appropriate knowledge and understanding are not operationalized at an institutional level.

Although there are islands of excellence in industry, education, and training following practices that demonstrate the adoption of green technologies, green jobs, and green skills, national consensus, strategy, and direction, are yet to evolve. While the government may push forward with reforms and the inclusion of green standards with a regulatory perspective, active interest of large corporations is important, since they can mobilize a countrywide movement to generate sustainable employment. The work of these corporations can also trickle down to small, medium, and micro enterprises. Also, the recent increase of financial and technical aid from foreign institutions (private, bilateral, and multilateral) also provides a positive environment that is conducive to introducing the notion of inclusive growth and the greening of the Sri Lankan economy as part of foreign aid and investment programs in the country.

While the Government of Sri Lanka may proactively lead such efforts, and direct education policy and institutions could start including generic and specific green skills in their programs and courses, active engagement of these institutes with industry is more critical to ensure continuous feedback on curriculum reform that is relevant to industry. While demand may not be imminent for such skills, requirements are going to rise in the medium and long term. Sri Lanka should be proactive to such future needs.

Given the recent post-civil war development of Sri Lanka, which has seen the emergence of new, vibrant policies for economic growth, and the attracting of foreign aid and investment, the country is firmly on track to economic recovery. The impact of progressive and forward-looking policies generated by the government, the active participation of the international development community, and efforts toward achieving more social cohesion are all works-in-progress, and so the long-term impacts of such major initiatives are yet to be seen.

3.2 Summary of Main Findings

The Sri Lankan economy has achieved a rapid rate of growth over the past 5 years. During this period the country has been able to attract more foreign aid, private investment, and a $2.5 billion loan from the International Monetary Fund (IMF). However, the benefits of this rapid development have not been spread uniformly across all regions. Imbalances are observed in certain areas such that the Northern Region, Uva, North Central, and Sabaragamuva have a much smaller share in the overall GDP compared with Western Province, which contributes up to 44.4% toward the national GDP. The country has one of the highest literacy rates (up to 90% depending on which region of Sri Lanka) in Asia, which contributes to an increase in migratory workers and rising incomes. Nearly 61% of the population is below 34 years of age, and the dependency ratio is approximately 61%, posing a big demographic opportunity for the country (Economic and Social Statistics of Sri Lanka 2013).

The *Mahinda Chintana Vision Document 2010* of the Government of Sri Lanka envisions a transformation of the country into a strategically important economic center with a focus on human resource development and environmental protection. It estimates developing a manpower reserve of 350,000 skilled professionals. In addition, the Haritha (Green) Lanka Programme aims to mainstream environmental concerns into the national development planning process.

To stimulate greening of the economy, the Ministry of Environment (MOE) has established the National Green Reporting System with a view to promoting the reporting of sustainability performance in the manufacturing and services sectors. Various ministries have initiated programs related to environmental issues. Whether it is the *Divi Neguma*, a program for cottage industries; training the trainers programs by the Tertiary and Vocational Commission (TVEC) or by MOE; the setting up of institutions such as the Ceylon–German Technical Training Institute (CGTTI), supported by GTZ; or the Sri Lankan Institute of Advanced Technological Education under the Ministry of Higher Education, the intent of the government is progressive and reform oriented. Even international agencies such as ILO have initiated demonstration projects in Western Province in solid waste management.

The impact of progressive policies has resulted in the opening up of new employment opportunities that require specific technical knowledge and skills.

About 50,000–70,000 skilled people are needed annually for large-scale development activities planned in the areas of naval, aviation, commercial, energy, road and transport, urban, and irrigation development. The industry sector requires nearly 500,000 skilled workers, while the services sector needs approximately 1 million skilled workers over the next 5–6 years. However, with limited institutions, the government estimates that only 30% of these requirements can be met with the existing training infrastructure.

There are approximately 1000 institutions under government and statutory boards, a large part of the training (55%) being with the private sector and NGOs. While the government training landscape is divided among six primary agencies, policy formulation and planning, quality assurance, coordination, and the development of TVET are done by TVEC, which has developed 17 vocational education and training plans for key industry sectors and has established the Sector Policy Training Initiatives. If the curriculum goes beyond National Vocational Qualifications (NVQ) standards, other players such as Department of Technical Education and Training (DTET) develop the curriculum for general subjects, in consultation with industry. TVEC has also played a unifying role for all TVET efforts in the country and for achieving the broad goals stated in the national policies.

3.3 Summary of Key Findings from the ADB-EdUHK Surveys and Interviews

The ADB-EdUHK surveys and interviews captured the voices of important stakeholders from industry, TVET training providers, government, NGOs, and development agencies. A variety of issues were discussed ranging from the most effective ways to green the economy to the best ways for TVET providers to update their programs to become more relevant to meeting modern economic needs, including developing green skills for green jobs.

Absence of Green Skills Affects Business

Most respondents from the energy, construction, and transport sectors recognized the importance of green jobs and agreed that the absence of green skills is impacting their businesses. Though the response rate from the hospitality and apparel sectors was low on this issue, larger hotels reported that they had been able to mitigate the gap by providing in-house training. The same level of awareness and commitment to training was not evident in small and medium hotels. Transport companies reported a lack of green skills in many functional areas of their operations.

National Policies Affect Businesses, but Rising Costs are a Major Deterrent

While the transport sector is most affected by national policies, the construction and energy sectors are impacted by global environmental regulations and by pressure for developing sustainable infrastructure. The apparel sector is also guided largely by international mandates and policy guidelines. However, rising costs across sectors are putting pressure on businesses to adopt sustainable practices by using energy-efficient materials, though understanding about the use of sustainable infrastructure has been unusually low across sectors, indicating a poor understanding of the benefits and cost advantages of developing infrastructure that is environmentally friendly. This area needs significant capacity building across industry sectors.

Downstream Supply Chains have Low Expectations, but Upstream Supply Chain have High Compliance

Overall interest in the sustainability of the supply chain is generally low across sectors. Hospitality sector respondents indicated most interest, with 40% expecting suppliers to use sustainable infrastructure, largely influenced by supply chains in food and raw materials. The apparel sector, which maintains high upstream compliance for sustainable infrastructure, does not regard the same for the downstream supply chain. Fewer than 20% of respondents expected their supply chains to be energy efficient, to use sustainable infrastructure, or to use green products. Around 50% of respondents from the energy and transport sectors expected the supply chain to be energy efficient.

Businesses Lack Engagement with TVET Institutions

Representatives of all the business sectors reported that their engagement with TVET is low. Fifty-six percent of respondents from the energy sector and 40% of respondents from the hospitality sector indicated that they provided on-the-job training, whereas overall engagement with TVET remained weak. This low level of satisfaction has triggered major companies to develop their own training institutes.

However, TVET Institutions Report High Engagement with Businesses

However, the TVET sector provided a different response when asked about its engagement with industry. It is possible that the response of TVET institutions was based on industry-level interactions that are not necessarily within the same industry sectors covered in the ADB-EdUHK survey, and may include manufacturing or other sectors. National Apprenticeship and Industrial Training Authority (NAITA) institutions seemed to have the closest links with industry. For example, the Automobile Engineering Institute has links with more than 100 agencies and/or enterprises, such as Nissan and Mitsubishi, for workplace training. The NAITA institutions also reported having the most flexibility to adjust to industry needs and to bring about changes in curriculum. Although most TVET institutions agreed about having close linkages with industry, fewer than half of the Vocational Training Authority (VTA), DTET, NAITA, private, and NGO institutions reported being members of any industry group or an industry association.

Impact of Skills Development Policies High on TVET Institutions, with NAITA and UNIVOTEC Most Proactive

Policies on skills development have a greater influence on TVET institutions than do economic or environment policies. The highest influence is on NAITA and DTET institutions. NAITA and the University of Vocational Technology are most proactive in terms of bringing about changes in their programs to manage transitions toward a green economy.

Formal Demand for Green Skills from Employers is Absent

TVET providers believe that although employers do not require green skills, vocational training providers (VTPs) have difficulties in addressing the development of green skills. Fewer than 40% of VTAs and around 20% of DTET institutions, other ministries, private providers, and NGOs believe they need to develop comprehensive approaches toward greening of their programs. Due to the emerging nature of the subject of green jobs, and a lack of any specific policy or curriculum on green skills, institutions are integrating green skills with existing training courses and programs. Most training providers remain uncertain about the issue of green standards. Further, due to a lack of internal capacity and limited professional development of teachers in the area of green skills, the institutes depend largely on external collaboration and experts.

Sector Issues

Construction. Sri Lanka has an estimated 300,000 skilled construction workers, with fewer than 5% of them formally trained and certified for green skills. However, due to the fast growth of the sector, it is estimated that another million workers would be required (ICRA 2011: 16). The emergence of the Green Environmental Rating System for buildings and other environmental and regulatory conditions is leading to changes in job roles, especially with regard to knowledge, skills, and attitudes, and also to changes in processes and work systems. For example, lack of knowledge and skills, and prevailing attitudes lead to wastage of raw materials Proper training in this area could lead to significant savings of CO_2 emissions.

Although companies have been hiring university graduates and TVET trainees, it is significant that only 25% of construction sector respondents in the ADB-EdUHK survey believed that TVET institutions are able to meet industry requirements. Larger construction companies recognize the importance of TVET and suggest that TVET providers should arrange industrial placements for training in safety and environmental skills and to develop quality assurance awareness.

During the interviews, shortages of construction professionals such as service engineers; quantity surveyors; project managers; construction managers; and other technically skilled jobs such as electricians, plumbers, masons, and fabricators were consistently reported. Respondents also suggested the need for upgrading skills concerning new technologies. Other areas suggested for skills enhancement were training in the correct application of cement, plaster, and other surfaces to reduce waste and poor finishes, top-up training in new building technologies, and effective recycling on sites.

Energy. Growth in energy, especially in renewable energy, is expected to generate high levels of employment requiring green skills. The government has set a target of generating 20% of electricity through renewable sources by 2020, mainly wind, as there are strong monsoon corridors in the country. As a consequence, the demand for training is increasing with regard to the manufacturing of small turbines, and assembly and installation of large turbines and high-voltage connections to the grid. Solar energy skills are needed in various areas such as installation and maintenance, though the sector is still emerging.

Fifty-six percent of respondents in the ADB-EdUHK study agreed that changes in processes in their companies have occurred due to a "greening" of the business. Changes in job roles and work processes are also triggering a demand for newer skills. Only 33% of respondents in the energy sector agreed that TVET institutions have the ability to provide for industry requirements. Large companies in the sector commented that TVET graduates lack practical knowledge and that their courses are out of date.

The businesses are in favor of increasing informal and on-the-job training in the areas of installation and maintenance skills for wind turbines and the installation of solar panels for domestic use. At a higher skills level, energy companies require sustainable product development specialists, environmental managers to oversee project site development, International Organization for Standardization (ISO) auditors, and site environmental officers to maintain standards and compliance.

Transport. The dominance of road transport (nearly 93%) has given rise to issues related to freight and passenger transportation, fuel efficiency, and maintenance of vehicles. Railways, air transport, and sea transport have initiatives around waste disposal, energy efficiency, and green buildings leading to requirements in green skills but the scale of road transport and its related greening requirements need more attention.

Only one-third of transport companies agreed that they had brought about changes related to work systems to promote the greening of their businesses, and a smaller number had made changes in job roles, processes, or value chain requirements. Changes in work systems center primarily on the challenges of fuel efficiency, including the deployment of new technology for further fuel efficiency, pollution control, and alternative fuel options. Job changes are related to the introduction of liquid waste recycling, the removal of solid waste, and the separation and recycling of parts and components.

Respondents from the transport sector identified the need for solid and liquid waste management and disposal, particularly relating to spillages. The transport sector as a whole identified a shortage of environment managers, energy managers, managers for aviation fuel, and supervisors of aviation fuel management and flight operations. Understanding the safe usage of chemicals to comply with Registration,

Evaluation, Authorisation, and Restriction of Chemicals standards was another area identified as important.

Services. Though hospitality and tourism have been growing, there is a massive shortfall of services and facilities compared with the increasing number of tourists. Due to changing customer demand and international regulatory requirements for energy efficiency, almost 60% of enterprises in the ADB-EdUHK survey had experienced a change in work systems, while 50% had introduced changes in job roles. Some of the triggers for these changes were practices regarding conserving water, wastewater disposal, and conserving resources during delivery and transport.

However, sector employers continue to be concerned about the poor attitude of TVET graduates. Only 20% of respondents agreed that TVET institutions were meeting industry requirements. Respondents surveyed from large business enterprises identified the poor quality of work performed and recommended greater emphasis on soft skills, foreign language competencies, and personality development. They also believed that TVET graduates are better suited for operational level employment than managerial roles. While quality continues to be a serious issue, the output capacity in the country is also low. Against an anticipated requirement of 9000 employees per year, the country produces only 1500 graduates every year (Sri Lanka Tourism sector summary by Capital Alliance).

Large hotels require specific policies and practices to address sustainability and compliance, requiring knowledge of energy-saving working methods to reduce the carbon footprint, and skills to implement and/or maintain ISO certifications. In rural areas, hotels need skills to work closely and effectively with local communities to develop the means for promoting and sustaining local livelihoods. They also require green procurement skills, and cultural and environmental protection knowledge.

Manufacturing. Textile and apparel account for nearly one-third of employment in manufacturing, and nearly 65% of the workforce is reported to be skilled. The manufacturing sector as a whole accounts for about 52% of total exports from the country. Environmental issues are given importance, and women are employed at all levels in the industry. The textile and apparel sector provides the best employment opportunities in the entire country for women. Rural women prefer this sector, as it provides employment at their doorstep.

Nearly 50% of apparel industry respondents agreed they experience changes in work systems due to the greening of their businesses. Some of these changes may be triggered due to the "Garments without Guilt" initiative launched by Sri Lanka's apparel sector to focus on ethical manufacturing and sustainable development.

As the industry expands and job roles change, dependence on TVET institutions grows greatly. Large companies recruit graduates from institutions such as the CGTTI but have in-house training systems, such as the Hirdaramani Group or Brandix Eco-centre, training and mentoring in the areas of energy consumption, water consumption, and waste generation. SMEs generally recruit from technical colleges, and microenterprises prefer to train informally through on-the-job training. The industry also depends highly on trained women from local villages.

3.4 Recommendations

With the changing needs of industry where jobs are becoming more modular in nature, the training for such jobs also needs to become nonlinear or modular. Multidimensional skills, especially greening skills, help in reducing the cost of production, improving productivity, and maintaining high standards with respect to sustainable practices. Policy makers and practitioners have a major task to balance emerging needs and priorities with existing implementation frameworks.

Align Education and Environment Policies

The Government of Sri Lanka's commitment toward green growth is evident through various policy documents, especially the *Mahinda Chintana Vision—2010*. However, the lack of an adequate and competent workforce with appropriate skills remains a bottleneck. There is an urgent need to revisit standard approaches to addressing issues within education and training, and to align these with sustainable growth. Ensuring a transition toward green growth, with adequately trained human capital, is not easy to achieve. While the vision document is a first step to achieving a paradigm shift toward green skills, achieving consensus; coordinating among political, social, and industrial leadership groups; and aligning economic, industrial, environment, education, and skills development policies are important areas to be addressed. This requires bringing in international expertise and good practices to learn the best ways and methods of policy rationalization, and the subsequent capacity development of institutions and individuals concerning processes and tools to sustain the policy paradigm.

The key responsibility lies with the Ministry of Youth Affairs and Skills Development in close collaboration with Ministry of Industry and Commerce (MOIC), Ministry of Economic Development, and MOE.

Provide Incentives to Promote Green Skills

Incentives for the private sector and TVET institutions are important to motivate them to move toward green skills development. Enterprises offering green skills development training should be entitled to general incentives concerning vocational training under governmental regulations or be allowed to recover training expenses through users or sponsoring institutions. The government may encourage financial incentives (tax breaks) to promote green industries, which can subsequently generate green jobs and a need for green skills. The government may also provide incentives in the form of land provision, tax breaks, and the provision of training equipment to enterprises promoting green skills.

The key responsibility is with the Ministry of Finance and Planning along with MOIC.

Promote Greening as a Business Issue

Green skills is not a priority for business operations, because the pressure of achieving business targets, and stiff competition, push the issues of sustainability low on the list. It, therefore, becomes very important that greening should become a business issue, which can be championed by larger companies. The benefits of upskilling the existing workforce need to be communicated widely in the SME sector.

The key responsibility rests with large corporations such as the HOLCIM Group, Aitken Spence, Haleys, and Sri Lanka Telecom, or public sector corporations such as the Ceylon Petroleum Corporation and Sri Lanka Ports Authority.

Promote Green Line of Credits from Foreign Aid and Investments

There are large foreign investments by multinational corporations, IMF, and others in Sri Lanka. ILO has initiated training programs for VTPs, GTZ is leading vocational training programs in northern Sri Lanka, and ADB has a Skills Sector Development Program. The agencies need more coordinated action to instill the importance of transition toward sustainable practices and requirements for green skills development.

Aid agencies may promote documentation and dissemination of good practices (national and international) and also support models of innovation toward green practices and green education with private sector agencies and TVET institutions, respectively.

The key responsibility is with the aid agencies like ADB, the World Bank, and the UN, and with multinational corporations.

Include Green Skills as Part of Curricula in All Levels of Education

Knowledge about green skills need to be introduced into the existing curricula, not only in TVET institutions but also in schools and higher education institutions. Generic skills in training can have a progressive path to specific skills and courses. Research in higher education, therefore, becomes critical to develop newer practices and technologies that are environmentally sustainable. Higher education institutions may also take the lead in the professional development of trainers through in-service training.

While institutional initiatives are quicker to implement, large-scale reform such as alignment among NVQ levels and National Competency Standards is also required in parallel.

The key responsibility is with TVEC along with NAITA and public and private TVET institutions.

Rationalize Private and Government Training Systems

The state should create a framework to accredit and recognize the different forms of vocational training undertaken by different enterprises. Training in the private sector needs to be integrated with the formal TVET system. Together with efforts to strengthen the formal TVET system, policy makers should not ignore the existing private sector training system and should take appropriate steps to achieve parity of reforms in the two systems. Teachers working in private sector vocational training establishments (with the same competencies and capabilities) should enjoy the same incentives as teachers in government vocational training institutes.

The key responsibility lies with TVEC along with NAITA, DTET, the University of Vocational Technology, and private TVET institutions.

Involve Media

In addition to formal TVET, sensitivity toward the environment is often developed through social interaction and public communications. Green skills development should be understood not only as a matter of curriculum, and in terms of content, limited to TVET institutions, but as something that should be built into a larger ecosystem of education, workplaces, social interaction, and political debates.

All forms of media—print, radio, television, and social—can be encouraged to position messages so that the public sees the importance of society moving toward greener practices—with green skills development becoming a necessity rather than an option. NGOs and SHGs can be targeted with special green skills modules to reach out to the community and to involve women and marginalized groups, to encourage them to develop green skills. These may or may not be employment-related skills.

The key responsibility lies with TVEC and the National Environmental Protection Council (NEPC), along with media companies and NGOs.

Expand the Scope of Green Jobs and Green Skills to the Informal and the Rural Sectors

Most industry and government engagement with respect to sustainable growth is limited to the formal and organized sector, and address mainly urban issues. Research and policy also need to generate evidence of the needs and situations of environmentally sustainable jobs and skills in the informal economy such as small-scale manufacturing, trading, art, and craftsmanship, etc., and related to work in the rural areas, especially agricultural processing and other nonfarm-related work.

The key responsibility is with multilateral agencies such as ADB, World Bank, and ILO along with research institutions and universities.

4 Viet Nam

4.1 Background and Overview

Viet Nam is one of the few countries proactively taking initiatives toward green growth, green jobs, and green skills. While the United Nations Industrial Development Organization (UNIDO) together with Ministry of Industry and Trade (MoIT) developed a "Green Industry" strategy for sustainable industrial development, GTZ together with the General Department of Vocational Training (GDVT) has developed a TVET reform project at GDVT, with a special focus on greening.

While government action may be triggered by global mandates and agreements, the industry motivation in Viet Nam is quite fragmented and lacks a unified approach, mainly because of the large number of SMEs. Also, the different growth trajectories of each industry sector impede concerted effort toward sustainable practices. Environmental issues, therefore, move low on the priority list of some of these sectors. Large corporations, on the other hand, are highly sensitive to the importance of sustainable practices and invest in technology and human skills as they realize their positive impacts on cost saving, brand goodwill, and edge over competition.

Even though policy and practice are gradually converging and there are examples in Viet Nam that demonstrate the power of going green, a larger society-wide impetus is still required to demonstrate the advantage of skills that are sustainable, employable, and remunerative over standard skills. The government may create incentive systems, exert regulatory pressures, and push reforms, but the real indicator of the country acknowledging green practices would be evident once small and medium industries start recognizing the value of green skills.

Government TVET institutions can easily start green skills courses through policy measures; however, absorbing those skills and providing better compensation in industry will remain a bigger challenge. While success can be achieved in organized businesses, the larger challenge would be small and microenterprises and the informal sector, which have significant shares in the overall employment market. The private TVET sector could easily align its courses toward green skills development when there is a demand from the industry as a whole.

While green jobs may have different shades of green across sectors—more in agriculture, less in traditional energy production, more in renewable energy production, and even less in industrial manufacturing—generic green skills development across all jobs is relatively easier to identify and implement. These skills may lead to specific and more technical green skills in the future. The country's leadership across government, the corporate sector, and civil society need to initiate steps toward generic green skills not only through formal education and training but through a larger social interaction embedded in workplaces, social discourses, and political debates.

4.2 Summary of Main Findings

Viet Nam Rises After a Difficult Past

Viet Nam is a prominent developing economy in Asia and is expected to become the world's 17th strongest economy by 2025. To reach its current level the country has over the years navigated important potential obstacles such as colonization, war, central planning, and a closed economy. It is a growing and dynamic economy that is set to achieve important economic reforms and poverty reduction goals. It has also made considerable progress in economic, social, and environmental aspects of sustainable development.

Viet Nam's progress has been augmented in recent times by the government's double agenda of restructuring state-owned enterprises and developing the labor market. From 2000 to 2013, Viet Nam's GDP growth averaged 6%, reaching an all-time high of almost 9% in December 2007 and a record low of 3% in March 2009. During 2000–2011, Viet Nam achieved significant improvements in human capital development owing to high economic growth. Manufacturing, IT, and various high-tech industries account for the largest share of the economy. In addition, oil production and tourism are also dominant industries.

The literacy rate of Viet Nam is a high 94%, which has contributed much to the country's economic growth. However, the required level of skills in the country is relatively low, with less than 4% of the labor force having any vocational training. Nearly 85% continue to work without formal training or qualifications. As a result, many become self-employed or engage in household work. Both education and employment reflect a lack of gender equity and pose a demographic challenge, since nearly one-third of the population is below 24 years of age.

Green Growth is High on the National Agenda

The *Viet Nam National Green Growth Strategy* of 2012 seeks to promote "green thinking" by aiming to reduce GHG emissions; increase clean and renewable energy; and promote the greening of production, lifestyles, and consumption. In addition, Viet Nam's progressive policies on climate change have been put in place to strengthen the capacity of society to cope with climate change.

In addition, the government has collaborated with many other countries for official development assistance projects, e.g., with Germany, Japan, and Luxembourg. The government has also linked up with various international organizations to facilitate the planning and implementation of various greening initiatives, such as GTZ, ILO, the Food and Agriculture Organization of the United Nations, UNIDO, the British Council, and the World Bank. The government has partnered with the EU to study consumer behavior, motivation, and challenges toward sustainable consumption. The government is also developing a methodology and tool kits for Get Green Viet Nam, a program to increase the share of sustainable consumption by Vietnamese consumers and increase the capacity of consumer organizations and government for convincing and supporting consumers in choosing more sustainable consumption behavior. Many NGOs have also initiated greening measures in the country, including the Viet Nam Cleaner Production Centre, Viet Nam Business Council for Sustainable Development (VBCSD), Viet Nam Vocational and Training Association, and Viet Nam Environmental Industry Association (VEIA).

Education and Skills Development Need to Speed up to Match Economic Growth

The Strategy of Education Development 2011–2020 and *The Strategy of Vocational Training Development 2011–2020* both emphasize that vocational training is the joint responsibility of the state and the private sector. Quality accreditation for vocational training has been officially defined in the *Law on Vocational Training 2006*, and accreditation activities commenced in 2008. The new Education and TVET Acts of 2005 and 2006, respectively, provide a higher degree of autonomy to TVET institutions in enrollment, recruitment, program planning, and financial management, and lesser control over assessment and certification processes. This flexibility encourages TVET providers to review their courses and to communicate with industry, seeking feedback that could be used to adapt training to employers' needs. However, even though the system has empowered TVET providers, many of them are still not able to do enough due to lack of funding and technical resources to renew machinery and equipment, improve curricula, and establish industry linkages.

Further, coordination problems between the Ministry of Labour, Invalids and Social Affairs (MOLISA) and the Ministry of Education and Training (MOET), and implementation hurdles at the provincial level have led to a lack of convergence at the TVET institution level and have affected the quality and consistency of TVET programs. Many TVET graduates remain unemployed, indicating a mismatch between skills taught and skills required for employment.

While at one end of the system, the government is introducing initiatives to mainstream vocational education and remove the stigma of low status associated with TVET, at the other end the system grapples with governance issues. For example, the introduction of 2–3 years of professional secondary and technical education (under MOET), and 1–3 year vocational training programs for students who have graduated from lower or upper secondary education, toward a vocational intermediate certificate or collegial diploma (under MOLISA), are encouraging steps; but the poor quality of teachers, faculty, and instructors, and the low industry relevance of traditional vocational courses, remain critical issues.

4.3 Key Findings from the ADB-EdUHK Survey and Interviews

Since 1986, Viet Nam has undergone economic reforms and changes that have transformed the country from being a centrally planned to a market-based socialist economy. In addition, Viet Nam's decision to join the World Trade Organization has significantly shaped its industrial structure and has paved the way for trade partnerships with strategic partners, such as the US, EU, and Japan. The Enterprise Laws of 2000 and 2005, which aim at providing a level playing field for all enterprises regardless of ownership status, have also contributed to the progress and the development of a positive industrial and trade environment in the country. The government intends to implement an industrial policy for sustaining growth in various sectors such as agro-industry, food, machinery, mechanical engineering, and construction materials; and by developing key industries such as energy, chemicals, metallurgy, and mechanical engineering.

Adopting Sustainable Practices can Control Increasing Costs

The demand for green skills in Viet Nam is increasing in the five high-growth sectors of construction, transport, energy, hospitality and tourism, and manufacturing. The highest demand is in hospitality and tourism, followed by the manufacturing sector. There is a strong realization among industries that the best way to manage increasing costs is to adopt sustainable practices; however, adopting sustainable practices adds to increasing costs. Companies, therefore, need to manage the matters of meeting customer demands for green products and introducing green processes, technology, and skills. Many of the green skills demands in hospitality and manufacturing arise out of the need for impact assessments. The enterprises also expect high sustainability standards from their supply chain.

Regulatory Pressure and Competition are Driving the Demand for Green Jobs and Green Skills

Unlike other countries, industry in Viet Nam is driven actively by the skills development policies of the government, in addition to environment policies and economic policies. The energy sector has reported the maximum influence of skills development policies on the sector, and hospitality companies reported higher influence of economic policies than skills development policies. Competitive pressures are also affecting the transport and manufacturing sectors and are gradually triggering a demand for environmentally sustainable practices.

Industry and TVET Linkages are Poor

About 70% of all businesses view TVET training of potential employees as important, but only 10% of businesses actively engage and dialogue with training institutes. TVET providers are also very much aware of the changing scenario where the economy is going green and so have introduced new and relevant courses in their curricula, particularly in the last 2 years. However, in the absence of appropriate greening standards and certifications, the accreditation of such programs becomes difficult. In addition, a lack of practical experience and low pedagogical ability among teachers and trainers in green skills creates bottlenecks in implementation. Vocational secondary schools (VSSs) and vocational training centers (VTCs) are actively involved in staff development and the training of trainers.

Sector Issues

Construction. Construction for housing, transportation, energy, water supply, drainage, and waste disposal is an integral part of the National Strategy for Green Growth for 2011–2020 and documents released for Vision to 2050. The policy emphasizes construction in urban centers above Grade II standards.

The changing preferences of domestic consumers for green buildings are putting pressure on developers and construction companies to develop low cost, sustainable infrastructure options for consumers. Even though more than 50% of businesses agreed to introduce resource-saving measures in electricity consumption and water use, and there is a rising demand for green skills, the lack of availability of such skills has not greatly affected business. Lack of green skills has impacted only 23% of businesses, and only 18% believe their growth was restricted due to an absence of such skills.

Some of the skills articulated by industry concern the designing of green buildings, waste management, and pollution control. Demand is increasing for skills in reducing topsoil damage, building retaining walls for roads and bridges that can withstand flooding, installing sensors for cooling and heating, and energy-efficient ventilation systems.

Transport. Road construction and rehabilitation have been the main focus of most transport policies in Viet Nam in recent years. Rising levels of motor vehicle use, urbanization of road space, air pollution, urban congestion, and exposure to major flooding and storm damage have impacted the environment adversely, and therefore new regulations like high vehicle tax, transport planning compliance with environmental protection standards, and emission and noise standards have been put in place.

A large number of respondents from the transport sector did not see much impact of social demand (less than 10%), or a need for impact assessments (only 25%), leading to a requirement for green skills; however, emission norms, and a knowledge of environmentally friendly engines, are more likely drivers for this demand.

While active steps have been taken to train the large workforce expected in the sector (630,000 by 2020), only 30–35% are expected to receive training in the interim. This is due largely to a shortage of infrastructure and an absence of structured programs for the mostly unorganized transport sector.

The skills required for higher level jobs include understanding of a higher level knowledge of fuel-efficient technologies and carbon emission; and sustainable urban transport planning skills to balance the transport system among rail freight, air, sea, and road transport. The medium- to low-level jobs require skills in the repair, operation, and maintenance of energy-efficient engines.

Energy. To ensure energy security against growing demand, the country strategy is targeted toward enhancing national capacity in research and the development of technology to use clean and renewable energy, increasing international involvement in developing and using clean and renewable energy, and increasing clean and renewable energy ratios with respect to total national energy use to protect natural resources and minimize GHG emissions.

The sector is witnessing rapid changes with respect to new work systems, including job role changes specific to attitudes, technical knowledge, and specialized skills. The pressure for sustainable practices in the downstream supply chain is also high. Almost 80% of respondents agreed to the importance of energy efficiency requirements, and sustainable consumption and production processes from downstream supply chains.

Higher level skills are required in waste management and treatment, containment of site pollutants, environmental risk management, and determining costs and skills in replacing obsolete technologies. Skills are also required in collecting data to provide accurate measures, and for communicating and reporting to meet environmental standards. Lower level skills are needed in operation and maintenance, pollution control, and waste management.

Services. Nearly 870,000 workers will be required in the tourism sector by 2020, of whom the trained resources are expected to be about 58%, with 33% having higher education qualifications. Rising CSR and the introduction of international or national standards such as Green Globe or Green Lotus are having an impact on the requirement for new skills in the sector. Most businesses believe that green skills development can be channeled through TVET institutions, since the demand is far greater than the supply. With more and more businesses adopting green practices in electricity consumption, water use, and recycling, the requirement for green skilled labor cannot be developed exclusively in-house or through on-the-job training programs.

An understanding of international standards such as Green Globe and the national Green Lotus, energy auditing, natural sciences for guides taking tourists around environmentally sensitive areas, maintaining and troubleshooting alternative energy supplies and grey water systems, maintaining solar panels, conversion of furnaces to biofuels, and maintenance of waste and sewage treatment facilities are some of the emerging skills across the spectrum of higher to lower order skills.

Manufacturing. Textiles, readymade clothes, leather, chemicals, agroprocessing, electrical, electronics, motor vehicles, and automotive parts are strong subsectors of the manufacturing sector that are experiencing rapid advancements in technology, competition, global trade, and skilled workers.

Skills gaps at higher job levels are reported in the design and monitoring of solid and hazardous waste management and the management of contaminated sites, lean manufacturing,[2] and environmental impact assessment. Skills gaps are also reported at medium and lower levels, especially in water treatment and handling hazardous materials.

4.4 Recommendations

Evidence shows that skills shortages may be impeding the transition to green growth in Viet Nam. Demand from the industry sector for jobs that are both environmentally sustainable and decent is on the rise. Initiatives that heighten awareness of sustainable practices and green employment should be developed. But TVET cannot work in isolation. TVET needs to align with economic policies and new climate change and industry requirements. There is an urgent need to revisit standard approaches to addressing issues within education and training, and to align these with sustainable growth. But ensuring transition toward green growth, with adequately trained human capital, is not easy to achieve in reality.

Closely Align Education, Economic, and Environment Policies

Consensus and coordination among political, social, and industrial leadership groups is evident through the National Green Growth Strategy 2012. However,

[2]Lean manufacturing (or lean operations) is a strategy for business improvement by producing better products and/or better services by using fewer resources. "Doing more with less and doing it better." (Goetsch and Davis 2009, pg. 508).

alignment among economic, industrial, environment, education, and skills development policies is still difficult to achieve.

Fortunately, there is no lack of knowledge about the subject. Knowledge from within and from other countries, and partnerships with international agencies are available to bring in new perspectives, new models, innovative practices, flexible funding, and high-quality research to develop individual and institutional capacities to deliver large-scale programs around green economy, green jobs, and green skills.

Government, with the assistance of multilateral agencies, can bring in international expertise and good practices in order to learn the best ways and methods of policy rationalization and capacity development of institutions and individuals.

The key responsibility lies with the Ministry of Planning and Investment (MPI) to lead other ministries such as MoIT, MOLISA, MOET, Ministry of Nature Resources and Environment, and Ministry of Construction along with ADB, World Bank, and ILO.

Provide Direct and Indirect Incentives to Promote Green Skills

Direct financial incentives in the form of tax breaks, subsidies, viability gap funding, and innovation funds for the private sector to move toward sustainable processes, thereby generating green jobs, are all important. It is also important for TVET institutions to be motivated to move toward green skills development. Enterprises offering green skills development training should be entitled to general incentives concerning vocational training under government regulations or be allowed to recover training expenses through users or sponsoring institutions. The government may further encourage indirect incentives by providing land, training equipment, international exposure visits, and industrial exposure for TVET staff.

The key responsibility is with the Ministry of Finance (MOF), along with MoIT, MOLISA, and MOET.

Promote Green and Sustainable Practices as a Key Business Issue

The pressure of achieving business targets and meeting stiff competition, often pushes the issues of sustainability down the priority list. Businesses fail to realize that sustainable practices can solve problems of increasing costs, provide an edge over competitors, and enable compliance with climate change mandates. Greening should become a key business issue that is championed by larger companies. The benefits of adopting sustainable practices and upskilling the existing workforce also need to be communicated widely and strongly among SMEs through supply chains.

The key responsibility rests with large corporations such as the Hoa Phat Group, Doji Gold and Gems Group, and Vinamilk, and public sector corporations such as PetroViet Nam and Vinacomin.

Coordinate Action Between Foreign Aid and Investment Agencies

UNIDO has created a policy framework for green industry in Viet Nam; GTZ is supporting the "TVET reform project" at GDVT, with a special focus on Greening TVET; and, ILO, ADB, and other agencies are also working in skills development in a variety of ways. A number of MNCs are also investing in business and trade.

Achieving coordinated action by the agencies can effectively and strongly push the importance of achieving transition toward sustainable practices and requirements for green skills development.

Documentation and dissemination of good practices (both national and international) can be supported by aid agencies, while MNCs may support models of innovation toward green practices and green education with private sector agencies and TVET institutions.

The key responsibility is with aid agencies such as ADB, the World Bank and UN, and with MNCs.

Encourage Green Skills Through Generic Content in All Forms of Education Curricula: School, Technical, and University

Generic skills in training can contribute a progressive path to specific skills and courses. Content related to sustainable practices and climate change issues can be introduced into existing curricula, not just in TVET institutions but also in school and higher education institutions. Research in higher education becomes critical to helping provide direct guidance toward newer practices and technologies that are environmentally sustainable. Higher education institutions may also take the lead in the professional development of trainers through in-service training.

While institutional-level initiatives are quicker to implement, large-scale reforms may also be initiated simultaneously. For example, the National Skills Standard Developing Committee under MOLISA can not only develop and establish green standards but also dovetail these with existing National Skills Qualification Certification levels and National Occupational Standards.

The key responsibility is with the National Skills Standard Developing Committee along with MOLISA and MOET.

Coordinate and Rationalize Private and Government Training Systems

Training by the private sector is potentially important and should not be ignored. While reforms in government TVET providers are championed by the line ministries, reforms in the private sector can be pushed through by industry and employers. Parity in the quality of training, incentives, governance, and industry relevance is required between the two systems. In addition, the state should create a framework to accredit and recognize the various forms of vocational training undertaken by different enterprises.

The key responsibility lies with MOLISA and MOET together with large private sector corporations.

Promote Greening as a Social Issue

In addition to formal TVET, sensitivity toward the environment is often developed through social interaction and public communications. Green skills development should be understood not only as a matter of curriculum, and in terms of content limited to TVET institutions, but as something that should be built into the larger ecosystem of education, workplaces, social interactions, and political debates.

All forms of media—print, radio, television, and social—can be encouraged to position convincing messages so that the community sees value in moving toward greener practices, with green skills development becoming a necessity rather than an option. NGOs and SHGs can be targeted with special green skills modules to reach out to involve women and marginalized groups, and to encourage them to develop green skills, which may or may not be employment-related skills.

The key responsibility is with GDVT under MOLISA and the Centre for Environmental Education and Communications.

Expand the Scope of Green Jobs and Green Skills to the Informal and Rural Sectors

Most engagement with industry and the government with respect to sustainable growth is limited to the formal, organized sector and addresses mainly urban issues. Research and policy also need to generate evidence concerning the needs and situation of environmentally sustainable jobs and skills in the informal economy. Activities include small-scale manufacturing, trading, and art and craftsmanship related to work in rural areas, especially with regard to agricultural processing and other nonfarm-related work.

The key responsibility rests with multilateral agencies such as ADB, the World Bank, and ILO, along with research institutions and universities.

Chapter 4
Case Study of a Private Sector Firm in Indonesia

Abstract Key issues, challenges, concerns, and prospects arising from a case study of the Astra Manufacturing Polytechnic in Jakarta, Indonesia, are presented, analysed, and discussed. A descriptive portrait is provided of the case study firm. Steps in the accreditation process are explained and discussed, as are partnership arrangements between Astra Manufacturing Polytechnic and government, including on-and off-the-job training activities, and the Meister programme for existing workers. Key lessons learnt from this case study are presented to provide a role model for the effective greening of business forms in Indonesia and as a guide for business firms throughout the region.

Keywords Private sector firms · Informal and rural sectors · Integration of green learning opportunities · On and off-the-job training opportunities · Meister program

By Belinda Smith, International private sector skills specialist

Successful partnerships between private sector industries and education and training institutions are needed to implement skills development for green jobs

1 Introduction

This case study was conducted at the Astra Manufacturing Polytechnic in Jakarta, Indonesia. It is an example of the successful implementation of skills development for green jobs. Data were collected using a variety of methods, including a site visit to make nonparticipant observations of the facilities available to students and teachers working at the polytechnic.

Several key issues may be identified as arising from this case study.

R. Maclean et al., *Education and Skills for Inclusive Growth, Green Jobs and the Greening of Economies in Asia*, Technical and Vocational Education and Training: Issues, Concerns and Prospects 27, https://doi.org/10.1007/978-981-10-6559-0_4

Courses take 3 years full-time to complete and graduating students achieve a diploma level III as a junior engineer. Short courses are also provided. Astra International provides 35% fully funded scholarships, and the other 65% heavily subsidized. Seventy percent of graduates get jobs within Astra International and 12% in other companies. Eighteen percent continue their studies at higher levels, while some start their own businesses.

Although there are no special programs for green skills, each subject has an integrated approach. Students learn to comply with environmental regulations and standards. All teachers have previously worked in factories and therefore have the required high-level skills to teach courses and pass on their knowledge of green practices in workshop learning activities. Programs consist of 65% practical skills development and 35% theory. The competency standards of programs are aligned to job opportunities within Astra International. During the final year of courses, generic skills and green skills such as waste management and energy and water efficiency are taught. The workplace component of the polytechnic model is not treated as a separate adjunct to the learning process. Learning through participation in work that is inseparable from classroom and workshop training is emphasized. The success of this approach is indicated by the high number (69%) of graduates who are considered immediately productive (work-ready).

Astra has developed initiatives to support local communities and government vocational secondary schools (VSSs) despite knowing that the delivery of TVET in schools is more costly than general academic subjects. Teachers from these schools are trained for 2–3 months, spending time, for example, in factories gaining knowledge about new standards, processes, and policies that are subsequently passed onto the students attending the courses. A wide gap has been found between the basic knowledge, skills, and abilities of graduates from urban areas to those from rural and remote areas. In addition, for graduates from VSSs, finding a job is difficult due to the lack of equipment and the academic nature of the training.

This case study demonstrates a successful partnership between the private sector and education and training institutions to implement skills development for green jobs.

Established in 1995 as the Federal Technic Academy, the Astra Manufacturing Polytechnic changed its name in 1999 under the sponsorship of Astra International, one of Indonesia's largest conglomerates. Most of the company's operations are focused on the Indonesian market. The Astra group has 192,895 employees in 187 affiliated companies and nine foundations with 6304 employees. The company is the largest automotive distributor and producer in Indonesia, partnering with Japanese vehicle manufacturers such as Toyota, Daihatsu, Honda, and Isuzu. Astra International has six core businesses: automotive, financial services (mainly to finance automobile purchases), heavy equipment, agribusiness, information technology (IT), and infrastructure. Automotive manufacturing and distribution is the largest division. Astra Manufacturing Polytechnic is a technical institute in the suburbs of Jakarta, based within the Astra vehicle manufacturing compound. Green skills are incorporated across subjects in the curriculum with a focus on energy efficiency, waste reduction, liquid and solid waste management and disposal, recycling, water efficiency, and the reduction of chemical usage covered under the relevant areas in the training programs.

The polytechnic, also known as Polman Astra, is fully funded by the company through the Astra Bina Ilmu Foundation and is accredited as a polytechnic by MoEC. It conducts programs under five specializations

Astra Manufacturing Polytechnic at a glance

Company-funded, government-accredited

Situated in the company factory compound

82% graduate employment

18% graduates university

Strengthening six government vocational high schools by training government teachers to use CNG equipment

65% practical skills development

Integrated green skills development

670 students

Male 562 (84%)

Female 108 (16%)

2,072 alumni

55 full-time lecturers

:

- **Mechanical Engineering**, established in 1995 with Akademi Teknik Mesin Industri (ATMI) in Solo, a Jesuit polytechnic university;
- **Production Engineering and Manufacturing Process**, commenced in 1996 in cooperation with the Department of Mechanical Engineering at the Bandung Institute of Technology (ITB), a highly regarded Indonesian university;
- **Information Systems**, started in 1999 in cooperation with the Department of Information Technology, also at ITB;

- **Automotive Engineering**, established in 2003 in cooperation with Astra Automotive Group and ITB; and
- **Mechatronics**,[1] commenced in 2005, internally developed based on existing programs.

These specializations highlight the tradition of partnerships that Astra Manufacturing Polytechnic has established over the years. Several of these partnerships are ongoing. ATMI in Solo is the oldest and most established teaching factory in Indonesia with the best tool-making program in the country. The two training institutions work together to identify ways to improve the image of TVET in Indonesia. ATMI, Astra Manufacturing Polytechnic, and Bandung Manufacturing Polytechnic hold their own skills competition to improve student skills. The three organizations have different strengths from which each can benefit. ATMI has the most established teaching reputation; Astra Manufacturing Polytechnic has the industry linkages; and Bandung Manufacturing Polytechnic has strong linkages with the government. For example, while ATMI does not have strong industry linkages, it has alumni throughout Indonesia, many of whom are in senior positions and who employ graduates from ATMI and the other two institutions. If one of the three organizations has an excess of very good students applying for its courses, it will refer them to the other two training institutions. The institutions also support each other to improve their internal systems for training by sharing approaches and how to improve the model of TVET being delivered. They also share some of their training courses.

Astra Manufacturing Polytechnic is still developing and improving internal education systems. This is a problematic process, and different approaches are required. Astra Manufacturing Polytechnic first adopted a Swiss and German educational approach that has since been influenced by the more recent requirements of the Indonesian TVET system, which is influenced by Australian competency standards. This is overlaid with the Astra management system based on a Japanese approach. Sometimes Australian competency and Japanese standards have been problematic in implementation. This has resulted in reviewing and changing internal education systems to accommodate the best of each group of standards where possible.

The polytechnic sits organizationally under the Astra Bina Ilmu Foundation in the Astra Group. The foundation focuses on professional education in the field of technology, particularly in automotive and natural resources. Scholarships are provided to employees, polytechnic teachers, and instructors for associate, undergraduate, and degree programs.

Indonesian polytechnics were created in the 1980s and are usually linked to universities. Students are eligible for entry after successfully completing 12 years of education and completion of the Ujian Masuk Politeknik entrance examination. Polytechnics provide D II (diploma level 2) and D III level programs, which usually

---

[1]Mechatronics is a design process that includes a combination of mechanical engineering, electrical engineering, control engineering, and computer engineering.

involve 3 years of full-time study with pathways for graduates to continue on to higher education. Astra Manufacturing Polytechnic provides certification mainly at the D III level, which involves 3 years of full-time study plus 1 year of on-the-job training. Astra Manufacturing Polytechnic also provides short courses. To become an accredited polytechnic, the Astra Group sought accreditation through the National Accreditation Board for Higher Education (BAN-PT), which is under MoEC.

The accreditation process consists of the following steps:

(i) The institution conducts a self-evaluation and draws up a report.
(ii) Two or three assessors conduct a desk evaluation.
(iii) Accreditors conduct a site visit.
(iv) The assessors submit their report to the board of the BAN-PT for a final decision.
(v) Stakeholders are informed of the result of the accreditation.
(vi) Accreditors issue the accreditation certificate with recommendations for further development and program improvement.
(vii) The accreditation process is repeated every 5 years.

Along with the accreditation process, MoEC has education standards incorporating curriculum, competence, administration, financing, equipment, teaching staff, assessment standards, and minimum formal qualification levels for teaching staff. The commitment to maintaining and documenting the attainment of standards represents an additional effort to usual staff development activities. It also displays Astra International's commitment to provide formal recognition and access to educational pathways for their workforce. Astra Manufacturing Polytechnic considers the accreditation process and accompanying educational standards as a minimum set to be further reinforced with ISO quality and environmental standards. Both the students and the company believe the Astra certification is more beneficial and influential than government qualifications.

Demand for course places among prospective students is high, with an average ratio of 1:17. Subsidized students pay Rp12 million, while the actual course costs are Rp100 million. Seventy percent of graduates get jobs in Astra International; 12% in other companies; 18% continue their studies at higher levels; and some start their own businesses.

To gain entry into an Astra Manufacturing Polytechnic program, students need to take a series of psychometric tests. The tests do not emphasize academic ability. In addition to identifying potential graduates, these tests allow students disadvantaged by poorer schooling to compete with students from West Java. Students from East and West Java have a large gap in terms of academic performance. The quality of graduates generally declines from Jakarta in a radiating circle.[2] Astra Manufacturing Polytechnic is conscious of ensuring that they do not just recruit students from Jakarta and is working to ensure that students from remote areas are

[2]Information from interview discussions.

Table 1 Applicants and total intake, 2009–2012

Applicants	2009	2010	2011	2012
Total number of applicants	1153	1939	2006	3995
Scholarship applicants	240	1028	1076	2916
Regular applicants	913	911	930	1079
Total intakes	233	237	229	231

Source Official records

able to enroll. The polytechnic provides dormitories for students from remote areas and sometimes offers a matriculation program, yet it is still difficult for the students from the east to attain the same level as students from West Java (Table 1).

Astra certifies the student and acknowledges the factory-related specialized skill area of the student and his or her final project. The polytechnic does not undertake an advertising campaign to attract students, but consults with government TVET institutions and their factories. Astra Manufacturing Polytechnic has undertaken small labor market studies, but the training programs are essentially aligned to the demands of the various companies in the Astra Group.

In addition to the mandatory subjects of civics, religion, and state ideology required by MoEC, the Astra Group of companies develops the polytechnic curriculum and informs MoEC of its endorsement by industry. From the Astra Manufacturing Polytechnic viewpoint, the industry endorsement is particularly important, more so than a government endorsement, as recognition of student competency arises from the industry.

2 Integration of Green Learning Opportunities

Astra International is working toward becoming a green company and follows a number of policies to support this, including a green strategy, green process, green product, and green employee approach. The Astra Group wants to comply with all relevant environmental regulations, both national and international. These are incorporated into the Astra Manufacturing Polytechnic training programs and internships.

There is no special program for green skills. Each subject has an integrated approach to green skills, so as the students learn a set of technical skills, they are also learning to comply with environmental regulations and standards as well as the Astra Group's green policies. Since teachers come from the factories, they integrate green practices into the workshop learning activities. Astra Manufacturing Polytechnic supports the company's green policies, and the company practitioners provide insights to the teachers on new environmental requirements. The teachers then pass on this knowledge to the students. Green skills are incorporated across subjects in the curriculum in areas such as energy efficiency, waste reduction, liquid and solid waste management and disposal, recycling, water efficiency, and reduction of chemical usage. These are covered under the relevant areas in the training programs.

Additionally, many of the final improvement projects deal with ways to lessen the Astra Group's factories' impacts on the environment. In addition to an air pollution project, students have undertaken projects to reduce and manage oil wastage, which have led to changes being implemented permanently within the factory. Another project identified ways to increase material efficiency in the production process. This also led to changes on the factory floor. Other projects have included reducing waste, increasing energy efficiency, and improving the factories' cleaner production programs. There are 224 projects per year, many concerning environmental improvements. These environmental process improvements are implemented within the Astra Manufacturing Polytechnic training programs, ensuring that students are up-to-date with current work practices.

3 Partnership Between Astra Manufacturing Polytechnic and Government VSSs for CNG Mechanics

The MoEC Directorate General of Higher Education, which has a vocational schools unit, identified six schools with limited VSS capacity to participate in the Astra Manufacturing Polytechnic program, on the latter's request. The technical and training support is free to the VSSs and to the teachers receiving the training.

MoEC provided the VSSs with CNG machines in 2015, but only a few teachers were trained on the use of the equipment. Astra Manufacturing Polytechnic provided 3-days training to one or two teachers from each of the schools on using the CNG equipment. The teachers often send the polytechnic a proposal for additional training, and Astra Manufacturing Polytechnic provides further training if feasible. Astra Manufacturing Polytechnic has a memorandum of understanding with the schools to provide 2–4 training programs of varying length and content per year over 3 years.

Astra Manufacturing Polytechnic does not receive funding from the government for this program. MoEC has a similar program wherein polytechnics are requested to provide professional development to vocational teachers from vocational schools of a certain region. Astra Manufacturing Polytechnic considers its program to be complementary to MoEC's program and has extended it to create master schools. Schools located near the master school are invited to attend training, which is organized and implemented by Astra Manufacturing Polytechnic. The six VSSs identified by MoEC may become master schools to support other VSSs in improving their quality. So far, one master school has been established, and others outside Jakarta are being identified. Master schools will be chosen based on a long-term relationship with the polytechnic, the school culture, and teachers' abilities.

Successful implementation of TVET leading to meaningful outcomes for graduates poses a number of complex issues within a school environment. Some of these issues include teachers not having the skills and experience in the vocational areas where they provide skills development; the additional cost of equipment and facilities; and finally, access to resources so that students can practice and develop their competence. An ADB report[3] has identified that delivering TVET in schools is more costly than delivering general academic subjects. Astra Manufacturing Polytechnic has developed an initiative to support government VSSs despite the costs.

Astra Manufacturing Polytechnic formed the Centre for Vocational Education Development in 2011 to help government VSSs improve the quality and relevance of their student graduate outcomes. VSSs face challenges in preparing students for work. In most instances, the teachers have never worked in the industry for which they are preparing their students. It is very difficult for graduates from VSSs to find jobs,[4] due partly to the academic nature of the training and partly to lack of equipment. There is also a wide gap between the basic knowledge, skills, and abilities of students graduating from urban areas and those from rural and remote areas. This is further compounded by the East–West divide and the public–private school divide. One reason Astra Manufacturing Polytechnic decided to establish the Centre for Vocational Education Development is to redress the gap between students from different backgrounds. The vocational part of the VSS program is restricted to approximately 25% of the curriculum. This figure does not include mathematics, sciences, and English language, which are covered under the stipulated general education component.[5]

There is a major difference between general education foundation skills that might be technical in nature and the occupational competency standards in TVET, which are all about skills development within a workplace context (Smith 2012). Teaching general education technical skills is a common and traditional approach where education is contextualized to the world of work largely through incorporating narrow technical content into the curriculum, e.g., IT subjects or drafting skills where students learn to use perspective. The traditional education approach contextualizes technical skills in an educational environment. TVET demands skills that are contextualized in workplace conditions (deadlines, specifications, quality, etc.) and provide recognition for being able to perform that group of occupational skills to a particular identified industry standard. To develop occupational competence, students learn and practice applying vocational skills and knowledge in workplace contexts, under workplace conditions, including working with customers, working to a deadline and within a budget, and working under supervision. This critical change in emphasis requires a cultural shift within schools that

---

[3]ADB (2009). *Education in Focus. Good Practice in Technical, Vocational Education and Training.* Manila: ADB.

[4]www.unevoc.unesco.org/tvetipedia.0.html?tx_drwiki_pi1[keyword]=Indonesia.

[5]See Footnote 3 in Chap. 2.

recognizes industry relevance as one of the key indicators of quality in TVET programs. The Astra Polman Centre for Vocational Education Development program assists in either facilitating or bridging this cultural shift.

The programs conducted by the Centre for Vocational Education Development provide curriculum and learning resources and offer technical assistance and teacher training in soft and technical skills. They provide a range of courses and activities for vocational secondary high school teachers. The training duration is generally 2–3 months depending on the program content, with some lasting one semester. Most training depends on what the teachers require. They are not set programs and vary according to the teachers' needs. Certain programs require the vocational school teachers to attend training at the polytechnic, while others bring polytechnic staff to train the teachers in their work environment.

Training is provided across a range of technical areas in areas such as automotive, quality control facilitation, pedagogy, and soft skills. They aim to improve the teaching and learning processes within VSSs. Sometimes the training is very basic, such as teaching teachers how to use CNG equipment. Teachers from outside Jakarta will stay for 3 weeks and attend training at Astra Manufacturing Polytechnic from 7:30 in the morning until 4:30 in the afternoon. The teachers will attend training in the classrooms and workshops, and shadow some of the polytechnic teachers in the workshops while they are teaching. During their time at Astra Manufacturing Polytechnic, the VSS teachers visit factories and observe how skills are applied and the contextual issues considered important, helping the teachers to appreciate the importance of production processes combined with technical skills.

4 Harmonization of On- and Off-the-Job Training Activities

Astra Manufacturing Polytechnic is situated on the grounds of the Astra International main company compound. The compound contains the Astra International head office, the automotive manufacturing factories, and the distribution center. The training approach is completely integrated with the workplace at all stages of the learning process, not relying on an up-front or back-end model of training and workplace exposure. The teachers at the polytechnic originate primarily from industry and return to the workplace on a regular basis depending on the workplace schedules. If they do not come from industry, they participate in regular workplace training similar to an apprenticeship training program. The teachers who teach the practical skills are from the factories. Some teachers work 4 days a week at one of the factories and teach 1 day a week in the polytechnic. These part-time teachers are usually the heads of departments within the polytechnic. During their 1-day of teaching, they spend the morning coaching full-time teachers and in the afternoons are with the students. Other teachers return to the factory once a month or when the factory is not too busy. Some very experienced teachers who work 4 days in the factories also provide professional development training for the full-time teachers.

MoEC stipulates that teachers must have a master's degree; however, Astra Manufacturing Polytechnic considers it more important that teachers have industry experience and are competent in the areas they are teaching. This demonstrates that the polytechnic understands the value of TVET; that is, practical skills are recognized to be as valuable as other skills, depending on the context. The value of being competent in a technical area is an essential base for teaching technical skills. The polytechnic analyzes the training needs of its teachers and can then bring technical specialists from the factories to develop the teachers' skills. The polytechnic develops the pedagogical skills of their teachers and encourages student-centered learning, problem-based learning approaches, and competency development. It also sends the teachers to university to get a master's degree.

Previously, equipment would be updated every 9 years, but now they are upgraded within 5 years, along with some of the regulations associated with the equipment. This has an impact on the skills and knowledge of the full-time teachers. Their professional development and associated licenses are also costly to maintain. Environmental regulations change almost every year or two, which is where the workplace-based teachers are a very important link. They understand the changes that are implemented and introduce them into the Astra Manufacturing Polytechnic programs.

The different divisions within Astra International provide detail on the competency standards that students are required to achieve. The competency standards are aligned to job opportunities within the Astra Group of companies. Astra Manufacturing Polytechnic then develops the curriculum to match competency standards, embedding basic knowledge such as the compulsory subjects of Bahasa Indonesia, religion, and English and during the final year incorporating generic skills and green skills (such as waste management and energy and water efficiency). The teachers all have workplace experience of varying degrees, updated regularly to ensure that workplace best practice approaches are incorporated into workshop training.

Students first visit the factories before they start their first semester of training to observe what is happening in the workplace. A teacher visits with them and explains what they are observing. The students continue to visit the factories for observations every semester. During these visits, they are encouraged through prompts to consider back in the workshops and classrooms what they have learned. The method in the first year is to observe particular approaches. In the second year, the students must be able to identify various aspects of the work, such as the processes that lead to cleaner production and zero waste or lead to environmental compliance. In the final year, they work to apply their technical, process, and soft skills in the workplace.

The model in Fig. 1, developed by Astra Manufacturing Polytechnic, presents the various aspects of competency that make up the curriculum, and the relationship between the polytechnic and the Astra Group of industries. Many training

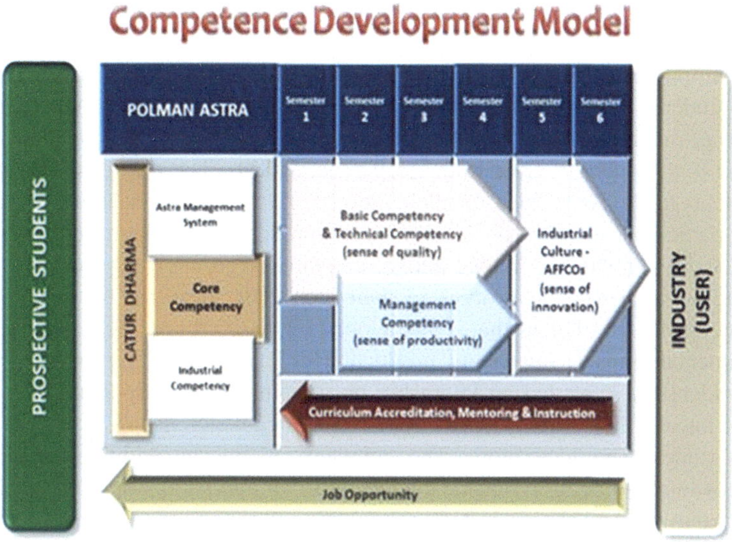

Fig. 1 Astra Manufacturing Polytechnic, 2013.

institutions in the Asia and Pacific region focus on a narrow range of technical skills development (Martinez-Fernandez and Powell 2009). Regionally, there is a recognized lack of soft skills and workplace process skills. Students may learn to weld two sheets of metal together but are not required to consider timelines, quality standards, or cost issues. In the Astra Manufacturing Polytechnic courses, programs are made up of 65% practical skills development and 35% theory.

The seven programs conducted at Astra Manufacturing Polytechnic are mechanical engineering and tool manufacturing, production process, mill, mechatronics, four wheelers, heavy equipment, and information management. These programs constitute the core competencies identified in Fig. 1 and are offered to meet an identified demand from within the different divisions in the Astra Group. The management competencies identified in Fig. 1 covers the Astra quality management system and include quality improvement. The industrial competency component represents the workplace internship. It is also a mandatory government requirement for polytechnics to teach religion, civics, and state ideology, which represent 1% of the curriculum. During the first year, students learn basic and technical competencies and develop a sense of quality through the processes of assimilating technical competencies within their workplace observations. From the second year, the program offers management competencies to develop a sense of productivity. Building on the technical skills developed in the first year, the second year leads to the accommodation of new learning (Piaget 1978) around the interplay of technical skills and productivity. In the final year, developing innovation skills is emphasized. Students undertake a 6–9 month internship or apprenticeship with one of the Astra affiliate companies as part of the learning process. This structure of learning is not only informed by pedagogical considerations, but also by the formalities of the workplace, where there is no room for less skilled workers attempting process changes or innovations until they have developed these skills.

The students are assigned a mentor who will teach the best practices in the workplace, including environmental compliance, and ensure that they are given exposure to a number of areas to develop general skills. The mentor will also help the students to devise an idea for their projects. Astra International encourages its staff to participate in "kaizen" (continuous improvement) through quality control circles. Staff members are divided into several circles (groups) and every year have to create an improvement process in their area. The ideas for the final project originate from a list of areas that are developed by these groups, including reducing environmental impact, improving production efficiency, and improving quality. Once students have started an internship or apprenticeship, they then choose two areas of interest. The quality control circles identify which aspect of one of these areas a student should explore. The company assigns an additional mentor if needed, who could be from outside the factory or an international person, to advise, support, and improve the project.

When the student completes the final project, the teacher and workplace mentor assess the project. If there is an external mentor, that person is also involved in the project assessment. Workplace learning is different from classroom and workshop learning in a number of fundamental respects. For example, workplace learning is usually focused on behavior, and workshop learning is usually focused on cognitive processes and the formation of skills. The aim of the internships is for students to apply their newly acquired skills and tools in authentic workplace environments, applying theory to their practice. Furthermore, students develop their technical skills within workplace processes and at the same time are complying with procedures and deadlines while cultivating the company culture. Opportunities are also provided for the factories to select and recruit candidates.

A final project must be completed by all students. It is not a theory-based project, but a practical quality improvement project. The aim is to develop innovation skills and reinforce skills in problem solving and continuous improvement. An example of a final improvement project is one in which a student worked to control air pollution in the automotive workshop by developing a pollutant reduction device. Another example involved improving the control system on a snap ring checker machine, which resulted in a decrease in the lead time and a reduction in the number of people required to undertake the task.

While mentors do not have formal training skills, the teachers engage with the mentors and monitor the students in the workplace at least once every 10 days or 2 weeks. This provides a formal structure to workplace learning, as they discuss student progress with the student and mentor, identifying any problems and discussing coping strategies. The teachers also review progress on the design of their final projects. There is a strong connection among the teachers, students, and mentors in the industry, providing a structure and creating a formal process for the workplace learning component. The company factories would like more interns; however, the polytechnic cannot meet their requirements, as they have a small student body.

The Astra Manufacturing Polytechnic model of integration for on- and off-the-job training is a sophisticated good-practice model. While the polytechnic has the advantage of being owned by the company, where students eventually find employment, useful insights are offered for all TVET training institutions on good workplace and institutional integration arrangements. The workplace training component is inherently pedagogical (Billett 2004), as students move from structured workplace observations of experienced practitioners at the commencement and in the early stages in the course, to learning as participatory practice (Billett 2004) in the latter part of the course. The course structure, as a whole, is intentionally sequenced to make the most of the complementary learning opportunities offered in the workplace and institutional context.

What is found in the Astra Manufacturing Polytechnic model is that the workplace component is not treated as a separate adjunct to the learning process, but rather that learning and participation in work are inseparable (Billett 2004). The success of this approach is indicated by the high number (69%) of graduates who are immediately productive (work-ready).

Meister Program for Existing Workers

Hager (1997) notes that one of the major impacts on the quality of learning that occurs in a workplace is the workplace environment or culture. The polytechnic has established a special 1-year Meister Program for existing workers, with entry requirements very different from other programs. Potential students must have high technical skills and up to 4 years of work experience. This entry requirement falls outside the guidelines of the Indonesian Qualifications Framework, which is not flexible enough to allow entry requirements of experience only. It is very difficult under current government regulations to recognize prior learning, so the solution is the Meister Program, which is accredited by Astra. The Astra Group of companies has many highly skilled workers who have learned their skills informally but cannot gain formal recognition, which this course provides. The polytechnic enrolls workers in the D III who have experience, but under current national arrangements cannot be given advanced recognition for skills and knowledge gained on the job. They can give concessions in some subjects to the experts, and so the Meister Program examines the skills of the workers against industry standards. If the standards are met, they will be given Meister recognition. While the program is not recognized by the formal TVET system in Indonesia, it is recognized by industry, as Astra is considered to have high standards. Workers, therefore, have a credible qualification should they change employment.

The polytechnic also conducts short courses for Astra workers that are open to the public, but mostly filled from within the Astra Group due to demand. These programs, combined with the company's commitment to the polytechnic, suggest that Astra International places a high value on education and training, which, in turn, reinforces the quality of learning that occurs in the workplace (Fig. 2).

At the time of placement, 69% of students are usually already able to meet the needs of Astra and contribute productively to the workplace. Students receive their certificates once they have finished all assessments and the apprenticeship. Some

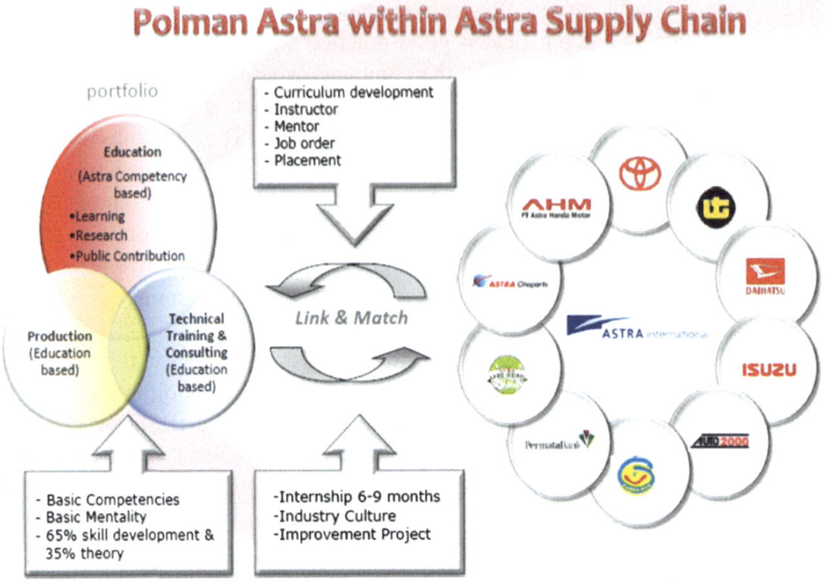

Fig. 2 Indonesia VET, 2012.

assessments occur in the workplace, particularly the final project. All assessments involve the teacher and the workplace mentor undertaking the assessment together, emphasizing the value of workers' expertise and their knowledge of applying technical skills to meet workplace standards.

Emphasizing the success of the polytechnic, several companies have visited Astra Manufacturing Polytechnic with a view to establishing their own polytechnics.

The Astra Group provides automotive technical manuals to technical vocational high schools throughout Indonesia. Assistance in the form of teaching and learning aids, including automobiles and transmissions, are also provided by the group.

Students from these VSSs are recruited onto Astra Manufacturing Polytechnic courses and receive a full scholarship to attend, which covers tuition and a small living allowance. Astra Manufacturing Polytechnic hopes that by improving the skills of students graduating from VSSs and entering the polytechnic, the overall standard of the polytechnic's graduates will also be higher.

5 Key Learning

All relevant environmental regulations, both national and international, are incorporated into Astra Manufacturing Polytechnic training programs and internships.

There is no special program for green skills, but each subject has an integrated approach to green skills. As the students learn a set of technical skills, they are also learning to comply with environmental regulations and standards, along with Astra Groups' green policies. Green skills are incorporated across subjects in the curriculum, with areas such as energy efficiency, waste reduction, liquid and solid waste management and disposal, recycling, water efficiency, and reduction of chemical usage covered under the relevant areas in the training programs.

In summary:

- First-year students learn basic competencies and technical competencies and develop a sense of quality through the processes of assimilating technical competencies with workplace observations.
- The second-year program offers management competencies to develop a sense of productivity; building on the technical skills developed in the first year, the second year leads to the accommodation of new learning (Piaget 1978) around the interplay of technical skills and productivity.
- In the final year, developing innovation skills is emphasized. Final-year students undertake a 6–9 month internship or apprenticeship with one of the Astra affiliate companies as part of the learning process.

In the Astra Manufacturing Polytechnic model the workplace component is not treated as a separate adjunct to the learning process. Instead, learning through participation in work is inseparable from classroom and workshop training. The success of this approach is indicated by the high number (69%) of graduates who are immediately productive (work-ready).

Astra International places a high value on education and training, which, in turn, reinforces the quality of learning that occurs in the workplace.

Chapter 5
A Holistic Approach to Greening TVET: A Case Study and Analysis of Bac Thang Long Economic Technical College Practices in Viet Nam

Abstract A word portrait is provided of the Bac Thang Long Economic Technical College with particular reference to short-term labor market forecasting, current long-term training programs, students' course preferences, monitoring internal quality, and relevance with particular regard to achieving internal and external efficiencies. The response of the technical college to challenges facing TVET in Viet Nam, greening practices, and implications of this case study to TVET providers and government policy and practice in Viet Nam are examined.

Keywords Short-term labor market forecasting · Students' preferences · Internal quality · Challenges facing TVET in Viet Nam · Greening practices · Soft skills

By Margarita Pavlova, International TVET Providers Specialist

Case study of successful approaches to developing skills in the context of transforming industries and the emergence of new green industries.

1 Introduction

International experience demonstrates that vocational training providers could play a significant leadership role in meeting the social and economic needs of greening. These training providers could integrate green concepts and processes into curricula and could implement green campuses, thereby preparing workers for new, changed, or emerging jobs in greening economies (NCWE and AED 2011). However, to stimulate the process of greening, current practices in curriculum development, links with industry, and patterns of training teachers need to be explored in order to provide targeted support for policy formulation and practice development in greening. This case study reveals these processes by analyzing a successful example of approaches toward provision of vocational training, including green skills development.

R. Maclean et al., *Education and Skills for Inclusive Growth, Green Jobs and the Greening of Economies in Asia*, Technical and Vocational Education and Training: Issues, Concerns and Prospects 27, https://doi.org/10.1007/978-981-10-6559-0_5

This case study was conducted at the Bac Thang Long Economic Technical College (BTL) in Ha Noi, Viet Nam and draws on multiple sources of information. It is an example of the successful implementation of skills development for green jobs, as set out in the TVET specialist deliverables. Data were collected through interviews with the college representatives, and secondary data were obtained from a variety of sources including college databases for student and staff records, and the college website.

This study suggests that current and future gaps in the green sectors (e.g., clean energy and green transport) are visible in Viet Nam. Many initiatives by aid organizations have produced a number of examples and approaches that need to be analyzed and implemented through formal TVET training. Current gaps in the *nongreen sectors* (the ones that have the most impact on greening economies) are not obvious, as these industry sectors are not "going green" at this stage. This research identifies future gaps in hospitality, construction, energy, and transport.

Directions for meeting the needs of the emerging green economy are sketched out in many government documents,[1] demonstrating the government's commitment and awareness of the issues relevant to green growth. Implementation of the Viet Nam National Green Growth Strategy, for example, will have a significant impact on TVET in the country. Solutions suggested in the document include education, awareness raising, and technical assistance to implement and enlarge production models based on respect for communities and the environment. Another area of attention stated in the document requires education and training provision for developing and running ecocities, green rural areas, green housing, waste management, and improvements in energy efficiency (Government of Viet Nam 2012a, b). This strategy requires an action plan where the role of TVET will be clearly stated. MPI has been developing the plan since the end of 2012. The quality, adequacy, and availability of green training provided at all levels should improve as a result of this plan. Currently at the low skills level, some greening initiatives are visible in the country (for example, pilot projects at craft villages, the green industry action plan for Hoi An City by UNIDO); however, formal TVET (vocational technical colleges [VTCs]) is not heavily involved.

Many strategies relevant to green growth adopted in Viet Nam are designed to meet the dual challenge of rapid economic growth and environmental issues, including climate change, and social issues. Collectively, they suggest a national strategy to first, adapt to climate change; second, to further enhance ecological sustainability; and third, to invest in science and modern technologies. These initiatives present long-term strategies for achieving key goals in the context of green inclusive growth. Bearing in mind government intentions in terms of a greening economy (for example, the Clean Production and Energy Efficiency Project, 2011–2016), gaps in the training for green collar occupations, in the short term, will be

[1]See, for example, Law on Environmental Protection (2005), National Strategy on Climate Change (2008), Green Industry Policy Framework (being drafted), Viet Nam Sustainable Development Strategy for 2011–2020 (2012), National Strategy on Environmental Protection to 2020 with Vision to 2030 (2012).

apparent. Targets such as an improvement in the productivity of raw materials, water, and energy; the minimization of waste and emissions; an increase in recycling and reuse; and the use of renewable energy (Viet Nam Government, draft) should be accompanied by skills development. Considering that a small number of institutions are able to start addressing these issues, and there are no specific plans in place, TVET in Viet Nam is ill-prepared to address the needs of industry with regard to the emerging green economy.

Bac Thang Long (BTL) has close links with local businesses to minimize the gap between TVET outcomes and job requirements. It uses well-established mechanisms to collect data through industry surveys, interviews, and workshops to ascertain the needs of industry and ensure that the content of courses is relevant to the skills required of students to undertake future employment.

Training programs are available for mid-level skilled occupations, and provisions are in place for students to transfer to university studies for high-level skills training. Student preferences and satisfaction are major influences when the college plans the content of long-term programs. Students receive workplace training as part of their courses, and their progress is regularly reviewed. They are fully supported by highly skilled teachers.

All courses are reviewed annually to ensure their relevance. For the training of existing workers, short-term programs have been specifically designed to meet industry needs. This is a good indication of quality training that is in demand by industry. Professional development is conducted regularly with teachers to ensure that their skills are up to date, and thus they are well qualified to teach students current theory and knowledge. Industry is fully involved in the professional development of teachers.

In anticipating greening of the economy, and in meeting the social demand for addressing environmental and social issues, BTL is proactive in introducing new courses and in effectively building awareness raising. Environmental legislation, which is the current attention of MOET, and the social responsibility mission of BTL, are the drivers behind these changes. The inclusion of generic green skills has not been addressed by the college, although it also pays keen attention to attitude development and takes measures to increase students' environmental awareness.

This case study concluded that effective practices in TVET provision observed at BTL could be scaled up in the context of green inclusive growth, as advocated by the government. Implementation of the Viet Nam National Green Growth Strategy will have a significant impact on TVET; therefore BTL could serve as an institution for driving forward green change. It could, for example, be a key, focal point member of a consortium[2] to assist with implementation of green skills best practices.

---

[2]International experience demonstrates that consortia are usually comprised of employers and industry organizations, training providers, and regulatory organizations. Funding is allocated by the government to those consortia that demonstrate best practices in green skills. These practices would then be implemented in sector-specific training.

2 Bac Thang Long Economic Technical College, Viet Nam

BTL is a public institution that started as a VTC that was built and supported by Bac Thang Long Industrial Park. In 2002, the Ha Noi People's Committee, and its Department of Education and Training, approved the transformation of the VTC into a semi-public college. In 2008, the institution changed its status to that of a public college. Now, it is an autonomous institution in terms of finance and human resource policies (such as personnel recruitment) and is located in Dong Anh, Ha Noi.

The experience of BTL is examined here as a case study of an innovative and interesting practice. In addition, many issues are revealed through this ADB project, for which the college is a microcosm in terms of issues, challenges, and prospects. The college experience is presented under two broad themes: existing and green practices. Existing practices ensure the relevance of skills development to the changing needs of the labor market in Viet Nam.

3 Relevance to the Society and Economy: Short-Term Labor Market Forecasting

TVET systems use different mechanisms to match skills supply and demand to reflect changes in the labor market and skills needs in their education and training programs. The approaches used could be implemented at different levels (such as the centralized–decentralized continuum) and could be holistic or fragmented. Approaches used by BTL are at the local level and are focused on short-term forecasting. The college uses qualitative needs assessment surveys and institutional dialogue (through a well-established and effective mechanism) to allow information about labor market needs and its translation flow into programs and courses. To achieve this, the college has close collaboration with local industry partners such as the Thang Long Industrial Park, Hoa Sao Media Group, Kim Cuong Media Corporation, Meling Plaza, Metro, Big C supermarkets, Van Tri Golf Company, Son Nam hotel, Ha Nam City, and Vinaphone. Annually (with the help of the Uniterra program[3]), the college conducts needs assessment surveys with local businesses to identify up-to-date information on the nature of companies, the composition of skills and qualifications of employees, their education, salaries, the

[3]The Uniterra is a volunteer program for Canadians to have the means to take concrete action and to make a real difference by applying their skills and experience. The Uniterra program is jointly run by World University Service of Canada and the Centre for International Studies and Cooperation. It is undertaken with the financial support of the Government of Canada provided through the Canadian International Development Agency. www.uniterra.ca/en/. The Uniterra program at the college helped to establish links with industry and to conduct needs assessment surveys.

ways companies advertise vacancies, approaches for recruitment, importance of workplace learning, professional qualifications, soft and technical skills for their future employees, specific technical skills required for junior level employees, relationships and channels of communication with TVET providers, quality of TVET graduates, and the training they organize for new employees. Information collected ensures that employers' views are well known to the college, helping it to plan relevant training programs and courses and to accommodate their requests in the college's plans, teaching, and training practices. This decentralized approach to labor market forecasting, together with MOET regulations and circulars (that translate policy decisions), are the main sources for planning new and adjusting existing programs.

4 Current Long-Term Training Programs

Currently, there are 12–15 long-term vocational training programs organized by three faculties:

Faculty of Engineering and Information Technology

- applied computer science;
- computer networking management;
- computer installation and repair;
- electronic engineering and telecommunication technology;
- telecommunication station installation;
- industrial and consumer electronics;
- industrial and consumer electricity;
- law; and
- office administration.

Accounting and Finance Faculty

- administration accounting;
- banking and finance; and
- business accounting.

Tourism and Commerce Faculty

- business in commerce and services;
- Chinese;
- cookery;
- English;
- Japanese; and
- tourism and hospitality.

Business Accounting and Telecommunication Station Installation are among the most popular programs. Electronic Engineering and Telecommunication Technology, and Industrial and Consumer Electronics are the least popular among students. The length of study depends on the educational background of enrolled students: 2 years for high school graduates, and 3 years for secondary school graduates. After completing college programs, students receive a TVET certificate at the intermediate level.

4.1 Students' Preferences

Although employers' needs are extremely important, meeting students' preferences is another factor that plays a significant role in determining the types of long-term programs and short-term training courses. This is particularly important because TVET has relatively low status in Viet Nam, with most school leavers preferring to enter university instead. Table 1 illustrates the key issues: the number of students in long-term vocational training has declined since 2008 (from 3779 to 1706 in 2013), when the college was established in its present format.

The number of students enrolled in short-term courses over the same period (2008–2013) increased from 700 to 4000. This shows the demand for specific, targeted skills development at different levels. The following short-term training courses, plus additional ones upon request, are available at the college:

- automotive technology;
- civil electronics and technology;
- foreign languages;
- high-tech welding;
- high-tech metal cutting;
- IT; and
- refrigeration and air conditioning.

Some short courses have been designed specifically for training existing workers, so the specifications for training required were provided to the college by industry. This is an indication of training appropriateness.

Student satisfaction, which is constantly monitored through surveys, plays an important role in the development and adjustment of courses. To attract additional students, BTL has established links with several universities such as the University of National Economy, Ha Noi Industrial University, University of Electricity, Senior College of Finance and Business Administration, and Senior Teaching College of Ha Noi.

This provides the opportunity to organize the articulation of college qualifications into higher education degrees, so students can transfer from the intermediate vocational training level (college) to a university (within 8 majors/training areas).

Table 1 Number of Students Enrolled at BTL, 2002/2003–2012/2013

	2002/2003	2003/2004	2004/2005	2005/2006	2006/2007	2007/2008	2008/2009	2009/2010	2010/2011	2011/2012	2012/2013
Vocational training	688	1850	2275	2175	2660	3779	4373	4782	4016	2752	1706
Transferring between levels of training (*From vocational training to college and/or university*)	x	x	x	102	338	753	1111	1433	1646	1879	1536
Short-term training (upon request)	x	x	x	x	500	700	1000	1158	2700	3765	4000

Source Data provided by Bac Thang Long Economic Technical College during the interview, 30 January 2013

BTL also provides the opportunity for students holding vocational certificates at the primary level to study for courses leading to an intermediate level TVET degree[4] (the process is governed by circular #55 on regulations for articulations). This collaboration attracts additional students to the institution. In the 2012/2013 academic year, 1536 students studied in those programs for level transfers (Table 1).

4.2 Ongoing Review of Courses

Among the college priorities is the ongoing annual review of courses to provide immediate responses to industry requests. For example, in 2012, some companies asked the college to improve studies in commercial trade by emphasizing learning by doing and spending more time on workplace learning. They requested that the college establish a model that combines student study and the practical application of skills, so theory and practice would be closely linked. The college is applying this suggestion in different ways: for a supermarket floor sales position, students learn and practice at the company; for accounting, students learn at the college and go to the company only for work placement. To increase the linkages between the college and industry, BTL asks company representatives to teach at the college. Around 30% of study hours at the college occurs with external support.

To maintain the relevance of its courses, the college collects very detailed information on industry needs. In addition to the surveys, it sends to partner industries, BTL interviews company representatives, holds workshops to clarify their needs, and then adjusts the curriculum to meet these needs. The rector has the authority to implement up to a 30% change within the curriculum framework developed by MOET. This approach of having 30% flexibility in curriculum development provides space for vocational training institutions in Viet Nam to engage in their own decision making about the curriculum. It also applies to institutions regulated by GDVT and MOLISA. As a result, programs and courses have been adapted and developed at BTL based on the needs of enterprises and on Viet Nam's education laws.

4.3 Internal Quality

Ongoing course reviews, as well as short-term labor market forecasting, ensure the external relevance to market needs of the training provided by BTL. Internal quality

---

[4]After their primary education, vocational training students study for 1.5 years to obtain a college degree (intermediate level); then they enroll in the second year of the university degree and study for 1.5 years to obtain a university degree (called a "college university degree"). This means that after 3 years of study a person with a primary vocational qualification can receive a higher education degree.

is achieved through the organizational structure of the college, approaches to teaching and learning (with a significant component of workplace learning), and quality of teaching staff.

The combination of training at BTL and workplace learning within local enterprises is the key factor that ensures high-quality training by the college. In addition to courses, reviews and workplace learning help ensure the quality and relevance of training. Practice is an important component of the curriculum. Two periods of workplace learning are included in the vocational training programs: 1 month after the first year and 3–4 months after the second year. These workplace learning placements are organized either by industry on the college premises or at the companies' sites. Ninety percent of students are enrolled in the college after high school (they study for 2 years) and 10% after lower secondary school (they study for 3 years). The availability of workplace training is an indicator of labor market relevance. During workplace learning, teachers from the college visit students and write reports on the students' practical experiences, wherein they list the tasks students are performing. The college then sends task lists back to the companies to evaluate and adjust as they see fit. These are then used as inputs in training improvements.

BTL has stated that "industry accepts the quality of [its] students." Although the college does not conduct tracer studies, it receives feedback from former students. Enterprises also recruit students from year 1 to work on weekends (for example, in restaurants and elsewhere in the hospitality sector).

Companies' involvement in developing training programs as discussed earlier in this case study ensures high-quality training through enterprise surveys commenting on students' skills and future requirements, feedback on students' performance during teachers' visits to workplaces, and participation in workshops that are attended by BTL and enterprise representatives to discuss college training programs.

Funding of the college comes from four main sources:

- 35% is from students' fees (around $70 per student per year);
- 30% from short courses;
- 5% from sponsorships (for example, by banks, industry); and
- 30% from enterprises for the short-term training of their workers.

Short courses are an important source of income and an effective way to achieve skills development or updates. The college-designed courses are based on signed contracts with a variety of organizations, for example, the Ha Noi Education and Training Department funding of D2 billion–D5 billion (up to $125,000) to train IT teachers. To overcome supply constraints, such as the inability of TVET providers to respond immediately to rapidly changing demands, the college applied for government funding (through projects) to support its infrastructure upgrading. Through these competitive grants, the government provides incentives to improve the flexibility the responsiveness of vocational training providers.

The college employs both part-time and full-time staff (Table 2) and adopts several measures to ensure the availability and guarantee the quality of teachers' professional development. It plans to engage 23 staff members over 2 years in professional development training to introduce new content identified through industry feedback and companies' needs surveys.

BTL combines both traditional training, such as when training programs are organized by MOET or the Ha Noi Department of Education and Training, wherein the trainer travels to the college, and more innovative training programs that are organized by international organizations. With regard to the latter, recently 10 teachers from Canada came to the college to train students and teachers. In addition, innovative ways and approaches to raising revenues for professional development include company-led training for teachers, the appointment of faculty heads from industry, and the establishment of the Centre for Training and Capacity Building.

BTL invites experts from various companies to work with every faculty member in the college, both at the college itself (the preferred option) and also outside (when related to students' practice placement). Teachers who are managing a group of students at a particular company are trained in the workplace and then evaluated by the company concerned using performance-based criteria. The heads of faculties are industry experts, but this is not true for all teachers working at the college. The Centre for Training and Capacity Building was established in 2012. Staff members from this unit regularly evaluate lesson plans and students' feedback, observe teaching practices, collect students' views, and seek to strengthen and upgrade the pedagogical capacity of College staff. The three main objectives of this unit are:

- Develop plans and implement surveys to evaluate teachers' performance, management quality, and to improve the quality of college human resources;
- Implement partnership projects for capacity building and skills development; and
- Organize a variety of training programs.

This new structure provides the opportunity to monitor and improve teaching and learning quality at BTL. The college also encourages teachers to link formal and nonformal learning, and recognizes the effectiveness of root-based initiatives in relation to professional development (PD).

BTL has a well-developed structure to mobilize its mission, which is to provide quality training programs including professional qualifications in a variety of fields based on frequent renewal of programs; and to ensure that training programs meet the various needs of students and employers. A number of departments support the work of faculties (Fig. 1).

Three centers established at the college also play important roles in supporting teaching, learning, and development. In addition to the Centre for Training and Capacity Building, the Accounting Practical Training and Counselling Centre and the Learning Resources Centre also support the work of the college.

The main purpose of the Accounting Practical Training and Counselling Centre is to

Table 2 Number of full- and part-time teachers employed by BTL

Full-time teachers			Part-time teachers		
Qualifications	No.	(%)	Qualifications	No.	(%)
Ph.D.	2	2.8	Ph.D.	14	5.86
M.A.	29	40.8	M.A.	50	20.90
B.A.	40	56.4	B.A.	175	73.20
Total	71	100.0	Total	239	100.00

Notes Total may not add up due to rounding
B.A. Bachelor of Arts, *M.A.* Master of Arts, *Ph.D.* Doctor of Philosophy
Source Data provided by Bac Thang Long Economic Technical College during the interview, 30 January 2013

Fig. 1 Faculties and Departments of BTL. *Source* BTL presentation during the interview

- develop programs to monitor, train, evaluate, and improve finance and accounting majors at BTL;
- organize internships and/or practicums for students;
- implement financial and account consultancy services upon request for accredited local enterprises; and
- monitor and evaluate the quality of teaching strategies and staff, as well as the qualifications of students.

The Learning Resources Centre holds multiple copies of more than 3000 books; copies of more than 30 different local magazines and newspapers; and a fully equipped electronic reference archive, including more than 100 reference CDs and tapes.

The close collaboration that occurs among all structural elements ensures the quality of teaching and learning, and students' and companies' satisfaction.

The college employs a variety of teaching methods such as student-centered learning, group work, and brainstorming, which are less traditional approaches in

TVET settings. Along with demonstrations and workplace learning, these approaches engage students in integrating theory and practice, with a focus on student-centered learning. Typical features of teaching and learning at the college are a mix of school-based learning; workplace learning; and extracurricular activities such as the managers' club, cultural performance club, sports club, and English club. This combination aims to develop appropriate knowledge, skills, and attitudes among students.

4.4 *Challenges Facing TVET in Viet Nam*

During the interview, a number of challenges confronting TVET in Viet Nam were identified by college representatives. The main challenge was the difficulty in attracting students. The overwhelming attitude in the country is to enroll at a university even though there may be no opportunity to apply the knowledge and skills gained at the tertiary level. In fact, it is not uncommon for university graduates to end up working as laborers. Universities are the main competitors for TVET, since students do not like the image of TVET, and regard TVET as low status. To overcome this problem, it could help break the TVET stereotype if the name of TVET institutions were changed to "institutes." An additional factor is that the salaries of TVET graduates are comparatively low, which makes TVET programs even less attractive to learners. Students' fees are an important component of the overall budget for many TVET providers, and because the fees are low, there are very limited opportunities for the providers to finance improvement of their infrastructure. As the providers' capacity is limited, the majority of TVET training in Viet Nam is outdated. Additional challenges are encountered due to the separation between MOLISA and MOET, with such management issues causing difficulties for colleges. Another external condition that has a significant impact on the colleges is the speed and direction of Viet Nam's economic development. The TVET system is vulnerable to in-country variations and to any external decline in economic development, such as during the 2008 global economic crisis.

These challenges are well known to BTL. They have been analyzed by the college and addressed through its policies and approaches toward training development and delivery, through articulation processes, through availability of short-term training, and through active positioning in terms of grant searching.

5 Greening Practices

Although the green growth concept is new to the college, it has been adopted in terms of both its social and environmental aspects that are relevant for the greening of TVET.

5.1 Social

In terms of the social aspects of green growth, the college has a strong commitment to social justice, paying special attention to providing access to studies for youth from lower socioeconomic backgrounds and in support of the local community. The college's mission statement highlights the objectives of providing educational and career opportunities that accommodate the aspirations and abilities of everyone, particularly youth facing financial hardships.

To achieve this objective, the college helps students to find jobs (85% of students work in local industries and study at the college); and to negotiate with employers (located mainly in the Bac Thang Long Industrial Park) concerning flexible working schedules that allow students to combine work and study. BTL develops flexible timetables for studies, with the same classes being scheduled both in the morning and in the evening. Students are paid around $150 each per month for their work, and the industry pays a fee directly to the institution of around D3 million ($150) per month. These guaranteed working places enable students from lower socioeconomic background families to enroll in training, to develop required skills, and to obtain qualifications.

Another approach used by the college to support poor students is through strong links with the local community. Through its targeted support of disadvantaged youth, the college sponsors 30 local students every year to train farmers to be employed in a range of jobs. In addition, college students generally rent accommodation in the local community, which is another means of providing community support.

5.2 Environmental

TVET in Viet Nam has sufficient flexibility in curriculum development that serves as an important factor in the greening of its curricula. As industry is becoming greener by responding to environmental regulations and laws,[5] so TVET providers can respond accordingly with regard to revisions in their curricula. Together with industry's requirements to address environmental legislation, the current attention of MOET to environmental issues is observable in the practices of BTL. These two forces behind such changes provide a good combination of specific and general, practical and more knowledge-based approaches, toward the introduction of green aspects into training.

[5]On 18 March 2013, the prime minister of Viet Nam signed Resolution No 35/NQ-CP on urgent issues in environmental protection that identified tasks to be addressed across different areas, including environmental protection in industry. A Green Industry Policy framework is under development by MoIT with support from UNIDO.

Some elements of greening have been included in the curriculum of several majors taught in the college. Three subjects in the electricity and electronics majors, the first two of which are compulsory for a number of programs, focus on environmental issues. All subjects are very new. This study has noted that TVET providers are very responsive to skills development policies; therefore MOET regulations and curriculum guidelines have the most influence on these green inclusions. Another driving force behind these developments is the Law on Environmental Protection (2005), approved by the National Assembly, which regulates key industries including energy. Article 33 refers to the need to develop clean and renewable energy and provides tax incentives to promote appropriate activities. In 2007, the government set targets to increase the share of renewable energy in total production from 3% in 2010 to 5% in 2020 and to 11% in 2050 (Decision No. 1885/2007/QD-TTg). There are also approved targets for wind and biomass renewable energy generation: from 3.5% of total electricity generation in 2010 to 4.5% in 2020 and to 6% in 2030 (Decision N. 1208/QD-TTg, 21 July, 2011).

The Programmable Logic Control (PLC) programming subject (automation), whose propose is to implement solutions to energy management and energy savings, was developed to address the regulations of MOET, to demonstrate practical applications of automatic control devices in industrial settings, and to address experts' recommendations at the time when faculty programs were developed. Through this subject, students develop their ability to design automatic control programs using processors that are widely used in industry. They need to learn about different components of processors and the ways they can be linked together to achieve a specified task. The content of the course, which was first offered in 2012, relates to energy saving and environmental protection, with regard to

- programming of automatic lighting control: simple thermostat equipment; automatic on/off control systems based on temperature and humidity;
- programming of automatic power switch systems: electricity, solar energy, and generator; and
- optimizing the use of fuel and raw materials: control concrete mixer models, elevator control models, and control waste treatment systems.

A second subject related to greening is energy efficiency use and energy saving, which aims to help students understand that the natural resources used for energy production are limited, the purpose being to change attitudes away from an excessive use of energy. This subject is based on the curriculum developed by MOET and reflects experts' advice at the stage of subject development. Its rationale relates to the need to save energy, increase energy efficiency, and reduce the costs of production and running a business. The college introduced this subject at the end of 2012. Objectives of the subject are

- to develop attitudes, awareness, and approaches toward using energy efficiently;
- to effectively apply knowledge about energy, forms of energy, mining technology, and energy use to real life situations and industrial production processes;
- to learn how to use energy-saving technology.

Content relevant to energy saving and environmental protection includes

- problems of saving and using energy efficiently in the world and in Viet Nam;
- new technology for energy saving;
- development of processes that save energy in households, agencies, and commercial government premises; and
- planning practices for improving energy use in lighting, refrigeration, and air conditioning equipment.

The third subject, use of renewable energy sources, is an elective of 100 h that was introduced in 2013. It is based on a curriculum developed by MOET from experts' recommendations, and aims to provide knowledge related to limited energy sources and the importance of the use of alternative sources of energy to protect the environment. The subject highlights the practical significance of the use of renewable energy sources to protect the environment and promote sustainable development.

Subject objectives are

- to differentiate renewable and nonrenewable energy sources and the long-term benefits of using renewable energy;
- to examine how to most effectively use energy-saving technologies and renewable energy instead of nonrenewable energy for environmental protection; and
- to develop appropriate attitudes, an awareness of environmental protection, and the priority use of renewable energy sources.

Content relevant to energy saving and environmental protection includes

- the concept of renewable and nonrenewable energies;
- forms of renewable energy (solar, wind, geothermal, wave, biomass, hydro) and nonrenewables (nuclear, coal, oil, gas);
- advantages and disadvantages between renewable and nonrenewable products; and
- a number of scientific and technological advances and practical applications of renewable energy use.

Two subjects relevant to greening are included in the tourism major in the Faculty of Tourism and Trading. Nutrition and food safety provides foundational knowledge on issues such as food safety and approaches to nutrition and food hygiene. Students are engaged in the analysis and evaluation of issues, so they develop an

Table 3 Subjects relevant to greening in the services sector (hospitality)

Number	Subjects	Duration (h)	Description
1.	Environmental protection	30	Provides students with basic knowledge related to the environment, and develops attitudes toward environmental protection in tourism
2.	Nutrition and food safety hygiene	45	Provides students with knowledge related to hygiene and food safety requirements in food processing processes
3.	Arrangements of work in the kitchen and kitchen techniques	45	Provides students with general knowledge about the whole process of labor arrangements; the processing of food and raw material supply in restaurants

Source Data provided by BTL during the interview, 30 January 2013

ability to design a nutritious menu and demonstrate, for example, knowledge of food preservation, and handling.

Another subject in the tourism major is environmental protection. This subject provides foundational knowledge on the environment in terms of habitat; the effect of habitat on human life and other life in the world; the need for environmental protection; analysis, detection, evaluation, and forecasting of environmental issues; and an ability to relate these in learning and to working life. After completion of this course, students are expected to have developed a capacity to improve the environment and prevent the adverse effects of environmental deterioration at their workplaces (be they offices, SMEs, or business establishments). They are taught to deal with issues that are related directly or indirectly to management, evaluation and environmental protection, exploitation and the use of natural resources, and economic activities.

Three relevant courses were introduced in the college 8 years ago in the hospitality major (Table 3). They were environmental protection, nutrition and food safety hygiene, and arrangements of work in the kitchen and kitchen techniques.

Tourism is recognized as an important sector for economic growth; however, it can also be potentially harmful to the environment. The college addresses a number of hospitality-specific green issues in its curriculum.

Future developments for increasing general awareness of environmental protection issues have been promoted by MOET. MOET, through a strategy to respond to climate change, which is supported by the EU, is in the process of introducing a 30-h subject on climate change at all institutions under its supervision. In terms of environmental protection, MOET has developed two separate subjects; one for agriculture (30 h) and another for industry (30 h). After February 2013, these became compulsory for secondary professional schools. There are 650,000 students at the intermediate TVET level who will be affected, and 19,000 teachers are to be trained (interview, MOET).

5.3 *Soft Skills*

Although green soft skills are not currently addressed at BTL, it recognizes the need to pay special attention to attitude development. Various strategies are used by the college to remind students of the need to think about environmental protection and to then act accordingly. On the home page of the college website is a green message encouraging students to save the power consumption of their computers by downloading and using a free software program called Granola.

The website states:

> Granola is free software that was launched to help save power consumption when using computers, but does not affect the performance of devices.
>
> This software uses a smart algorithm, combined with power management technology DVFS (dynamic voltage and frequency scaling) to help determine when to use the system at full capacity and when to simply use part of the hardware power. Granola will then automatically take steps to help optimize power savings when using the computer, but still ensures sufficient power supply for computer use when needed.
>
> This is like when you turn off the lights in the house when not in use to save power consumption, but you can still open them when needed.
>
> For light work such as web surfing, text editor …. Granola will help save up to 30% power consumption compared to conventional. When you use the computer to perform tasks like gaming or heavy graphics processing, the savings will be approximately 10% compared to the previous.
>
> Granola is also used effectively to help extend battery life when using the laptop. (BTL website)

Therefore, the students are encouraged to save energy. It is emphasized that the collective use of appropriate software results in very noticeable savings.

6 Conclusion

BTL is at an exciting stage of its development. The college combines several approaches to identify a demand side for skills, including short-term forecasting and up-to-date regular reviews of the skills required by companies, to plan and adjust training. Although it is training for middle-skilled occupations, it provides opportunities for students to develop their skills further to the level of high-skilled occupations. Mechanisms are in place that could put an additional focus on green skills to enhance their development in the programs. For example, when teachers are developing lists of skills for students to perform during workplace learning, they could identify (together with company representatives) current or potential green components to be included in training. This will potentially lead to a need to buy additional equipment, so the government could develop a grant scheme to support these initiatives by vocational training providers.

Environmental awareness has been included in the BTL curriculum and is recognized as a core skill required for all occupations. A number of developments have been formulated by MOET, and environmental regulations used by industry and the BTL mission, including social responsibility, provide a framework for "greening" the college. Other initiatives have been proposed by staff. The college intends, for example, to import solar panels to support their new environmental courses. Although the process of greening has started successfully by this training provider, environmental awareness has not been included across all courses. Generic green skills have not been formulated at the college or the ministry level, although a generic course on environmental protection will be introduced through MOET. Specific green skills are partly addressed through the health and safety courses. Values and attitudes development, through soft skills modules and extracurricular activities, are present across all countries that have taken part in this research. BTL has not included extracurricular activities that are related to greening (such as ecoclubs, a "green school" approach, or green activities in communities) or soft skills modules, although the college is aware of these possibilities. To support learning in a greener TVET curriculum, a systematic approach to teaching and learning methods that stimulates students' active learning and initiatives is required.

This case study has showcased the successful and beneficial relationships that can occur among a college, its students, enterprises, and the local community. The close and productive relationship with local enterprises, flexible training models, and meeting the needs of learners and labor market demands are key features of BTL's experience. The case study is also a good example of the significant, positive impact an institution can make on the development of local communities.

As a result of this case study, the following main points can be viewed as valuable takeaways:

- The government plays an important role in the development of TVET. The majority of providers follow government initiatives in skills development, and so the government needs to continue its direct involvement in training provision to increase its internal quality as well as its external relevance. Mainstreaming green skills in all skills development plans is important to achieve systematic reform.
- The introduction of new green courses should be related to a country's green development strategies and environmental laws and regulations.
- The content of vocational programs needs to meet the requirements of employers, students, and the wider economy. Currently, TVET does not use industry occupational standards for course development. Industry should be involved in initial curriculum development.
- TVET providers should have adequate freedom for curriculum adjustments that allow them to respond to industry demands regarding the greening of existing skills.
- To meet the requirements of greening strategies adopted by the government, TVET graduates need two sets of green skills: first, occupation-specific; and second, generic, soft skills that are related to building a general awareness of environmental protection.

- Greening should include social and environmental aspects that are closely related to the external efficiency of training.
- Systematic professional development, organized through structures such as the Centre for Training and Capacity Building, could ensure systematic training of staff and pedagogical development in relation to green skills.

TVET providers with good green practices should be identified as key players in consortia development by employers and industry organizations, training providers, and regulatory organizations. Government funding allocated to those consortia that demonstrate best practices in green skills could be scaled up to the sector-specific level of training.

Chapter 6
Summary, Conclusions, and Prospects

Abstract Critical themes in selected industry sectors (agriculture, manufacturing, building, transport and tourism) across the four case study countries are identified and discussed. A review also occurred regarding industry response to the demand for green skills; generic and specific green skills across industries of the four case study countries; the role of TVET in addressing industry needs; key impediments to the response of TVET to green skills; and, initiatives that work for countries to fast track reforms in TVET. Potential areas for further research are examined.

Keywords Regulation and sustainable practices · Industry response · Social demand · Consumer demand · Downstream activities · Upstream activities · Industry and TVET linkages · Demand for green products and services · Generic and specific green skills across industries · Green narratives · Role of TVET in addressing industry needs · Internal and external partnerships · Green content · Social demand · Areas for further research

1 Critical Themes in the Selected Industry Sectors

The pace of development in India, Indonesia, Sri Lanka, and Viet Nam demonstrates that growth and industry respond to the demands of the society and progressive policies rather than to different forms of government. While India is the largest democracy, the presidential democracy in Sri Lanka is being driven by a visionary leadership after a difficult, war-torn past. Viet Nam, on the other hand, had been a single-party socialist republic, while Indonesia has been dealing with relatively recent democracy. Despite such differences, in all the cases the pace of economic growth has exceeded expectations and has been due mainly to a positive response to the growth of the world economy, global trade, and high in-country domestic demand.

Industry sectors such as construction, energy, transportation, and hospitality services have been high growth sectors in all four countries. However, each of these sectors, despite their dynamism, faces constraints in the form of

R. Maclean et al., *Education and Skills for Inclusive Growth, Green Jobs and the Greening of Economies in Asia*, Technical and Vocational Education and Training: Issues, Concerns and Prospects 27, https://doi.org/10.1007/978-981-10-6559-0_6

- the informal nature of activities in the sectors,
- unorganized human employment,
- low penetration of new technologies and innovation in the various industries,
- poor implementation of policies related to environment in the sectors, and
- low productivity due to poor human capital development.

The ADB-EdUHK study has systematically evaluated the legislation, standards, and social demand in each of these sectors in the four countries, and has sought to understand the ways in which industry sectors are trying to respond to emerging needs and the key issues, including areas impeding a transition from traditional practices to green practices. The study has also examined various innovations in terms of policies and exemplar practices in both the government and private sectors.

2 Demand for Green Jobs and Industry Response

Unless otherwise specified, all figures are based on the EdUHK Team's analysis of data collected through the surveys and interviews undertaken for the ADB-EdUHK research study.

Strong Regulation Both Deter and Motivate Sustainable Practices

Rising regulatory requirements have an impact on all industry sectors, and many of the companies surveyed believed that being strongly regulated through government interventions was both a deterrent as well as an opportunity to instill environmentally sustainable practices. Forty-five percent of industry respondents in the ADB-EdUHK survey in India and 51% in Sri Lanka agreed that government regulation has an impact on their businesses (Fig. 1). Nearly 80% of construction sector and energy sector respondents in Indonesia also agreed with this proposition; however, respondents from services, transport, and manufacturing believed that

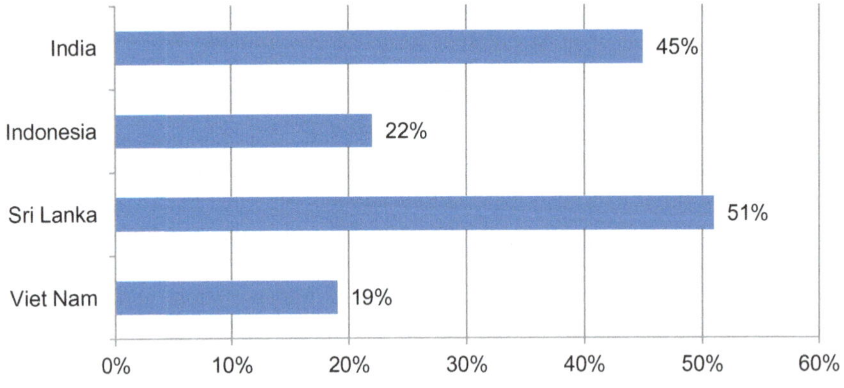

Fig. 1 Government legislation and regulations affect businesses. *Source* EdUHK Research Team Survey Questionnaires

other reasons, more than government regulation, have been influential, bringing the overall proportion in Indonesia to nearly 20%.

Some regulatory requirements, such as impact assessments, are the norm in larger enterprises where they have dedicated staff to manage such assessments. In smaller enterprises, these are usually outsourced, and ways and means are sought to cut corners in the process so that additional requirements do not impact the cost of production. Similarly, waste management is often a costly process in both large and small enterprises, forcing them to manage these matters more casually and cheaply. However, when enterprises come to realize the longer term benefits of such practices, find incentives for investment in practices to adopt green practices, and find skilled human resources to implement the programs, acceptability is much higher. This is the case in the larger enterprises and but less so in smaller ones. The adoption of such practices could give rise to green jobs and subsequently green skills.

While policy challenges exist for industries and businesses, such as a need to integrate economic, industrial, environmental, and skills development policies, institutional bottlenecks also exist to balance multiple efforts, ranging from compliance-related matters to ensuring productivity enhancement to saving costs, managing innovations to be competitive, and developing human capital.

Increasing Costs of Production Push Back the Importance of Sustainable Practices

Short-term targets often overshadow long-term outcomes, both in government and in the private sector. Rising costs, increased competition, and low profitability outweigh the efforts of technological innovations, skills development, and sustainable practices. Over 60% of respondents in Sri Lanka, and more than 40% of industry respondents in India, felt the adverse impact of rising costs (Fig. 2). Though the overall response rates in Indonesia and Viet Nam were relatively lower, sectoral response rates such as 70% in construction (in Indonesia) or approximately

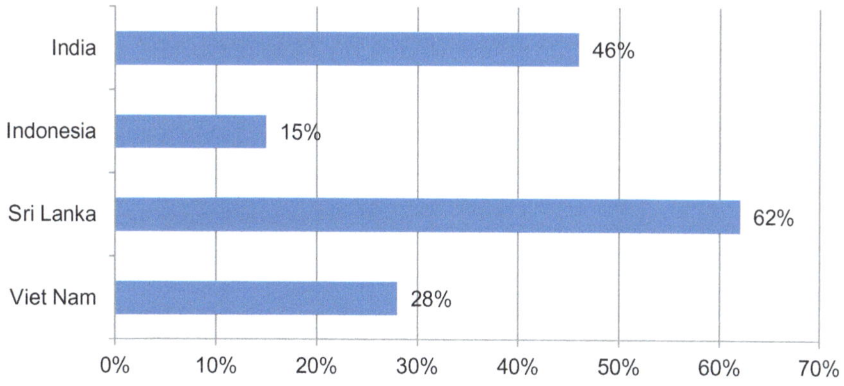

Fig. 2 Increasing costs affect businesses. *Source* EdUHK Research Team Survey Questionnaires

60% in manufacturing (in Viet Nam) confirm that the increasing costs of production are a serious issue.

Industries realize that the best way to manage increasing costs is to adopt sustainable practices. However, the dilemma persists that adopting sustainable practices adds up to increasing costs. While an institutional approach to sustainability has yet to evolve, individual motivation remains high among survey respondents across the four countries.

SMEs Have Less Motivation to Adopt Sustainable Practices; Larger Corporations are More Sensitive to Meeting Green Demands

The burden of existing issues often overshadows the importance of emerging issues in both public and private institutions. In addition, institutions are more responsive to complying with policy directions and regulatory provisions than they are to developing priorities for themselves. The urgency to deliver targets or exhaust budgets often results in institutions neglecting issues that may have a long-term impact and may go past a point of no return where no future action is likely to occur.

While it is difficult for government institutions to move out of the framework of their constitution to address emerging issues quickly, SMEs are focused primarily on achieving profits. Although they may superficially agree with emerging issues such as the green growth agenda, they may find it difficult to accommodate this in a highly competitive market. For example, almost 50% of respondents in Sri Lanka and in India agreed that they are influenced by industry standards and competition (Fig. 3); and 80% of construction and 100% of transport respondents in Indonesia felt the pressure of industry standards over their businesses.

Fortunately, every country has a combination of large public and private corporations where an awareness of the importance of sustainable practices and green skills is high. In addition, the idea that green skills are appropriate for all jobs is more widely understood than is the concept of green jobs, and companies that have

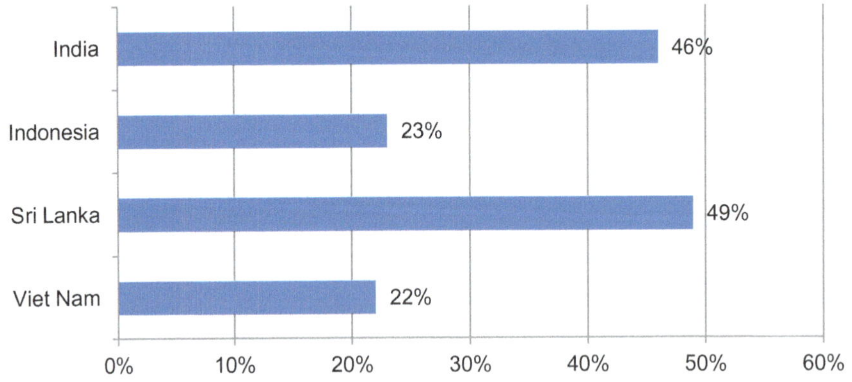

Fig. 3 Industry standards and competition affect businesses. *Source* EdUHK Research Team Survey Questionnaires

strong international linkages tend to be more appreciative of the importance of sustainable practices. They also possess international and national environmental accreditations of one kind or another, and value customer demands for green products, and compliance across supply chains. In many large corporations, CSR mandates and import–export requirements are key drivers toward developing green jobs and green skills.

Strong Social Demand and/or Customer Demand Is a Big Incentive for Encouraging Change Toward Sustainable Practices and the Requirement of Green Skills

National and international standards for products and services depend on the changing preferences and choices of consumers. Awareness about the involvement of child laborers in production activities has dissuaded many customers from patronizing products where child labor is involved. Sensitivity toward greener production has attracted customers to buy more "green products" even if more expensive. Such buying behavior, and pressure from rights-based groups, have influenced the structuring or restructuring of the standards set for different industry sectors. Thirty-four percent of respondents in India, 29% in Indonesia, and nearly 20% in Viet Nam and Indonesia regard rising consumer demand for green products and services as being important (Fig. 4).

In addition to changing customer demand, industry standards have also influenced global climate change mandates, competitive pressures, and government regulations, which get reflected in skill demands by industry. For example, cleaner technology processes in power plants have triggered the need for corresponding training and education. The introduction of CNG has highlighted the need for training about clean and green fuels.

While most industry respondents across industry sectors recognized the importance of green jobs, a smaller percentage agreed that the absence of green skills is

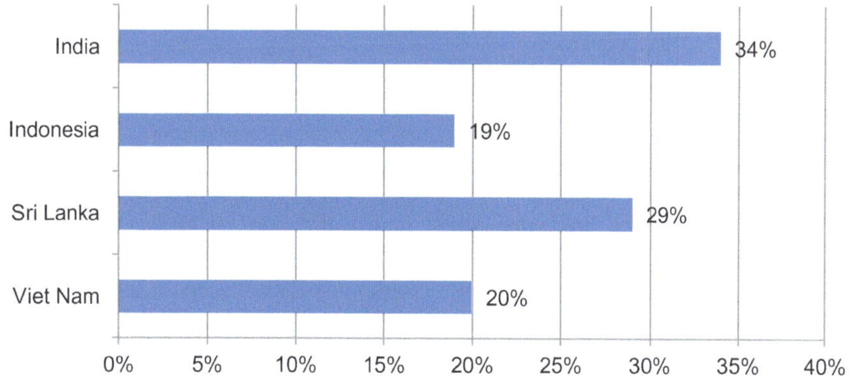

Fig. 4 Increasing demand for green products and services, standards, and competition affect businesses. *Source* EdUHK Research Team Survey Questionnaires

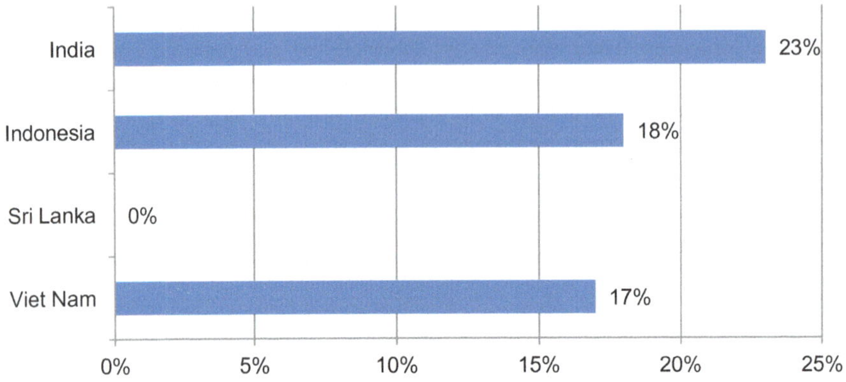

Fig. 5 Demand for green skills affect businesses. *Note* No data available from Sri Lanka. *Source* EdUHK Research Team Survey Questionnaires

impacting their businesses. Only 23% in India, 18% in Indonesia, and 17% of businesses in Viet Nam agreed that demand for green skills had an impact on their businesses (Fig. 5).

Expectations from Downstream Members of the Supply Chains Vary, but Compliance with Upstream Supply Chains Remains High

Industry sectors such as services (e.g., hospitality) and manufacturing (e.g., apparel), which have a strong dependence on raw materials from downstream members of the supply chain, also have strong expectations regarding sustainable practices from their vendors. This may also be because the customers are closer to industry and so there is a direct interface between the two parties. However, this situation is not as strong in other sectors, such as construction, energy, or transport, because supply chain products are less likely to have a direct interface with customers. When it comes to upstream supply chain mandates, enterprises across sectors are robust in managing international standards, client demands, and global trade treaties and mandates.

Industry and TVET Linkages are Weak and Cosmetic, Giving Rise to Parallel Training Provision

The low level of engagement between industry and TVET institutions continues to deter the improvement of courses and development of industry-relevant curricula. Although the nature of links may differ across the four countries, with a few institutions doing extremely well in terms of industry links (such as the Central Institute of Plastics Engineering and Technology Micro Small and Medium Enterprises [CIPET, MSME] tool room in India, or the University of Vocational Technology or NAITA in Sri Lanka, or some of the VSSs in Indonesia), the majority of training providers work in isolation and continue to do traditional training as mandated by archaic government content and curricula. Often the faculty designing and teaching courses remain complacent and not updated about new technologies and developments in the industry concerned.

As a result, several large, formal corporations have provision for in-house training in order to develop customized training programs for employees to facilitate a smooth transition into their respective jobs. For example, Infosys Technologies in India has invested more than $450 million in building its own training center. Wipro Technologies in India allocates 1% of its revenue every year to the specific purpose of training fresh graduates (Technopark 2009). The Hayleys group in Sri Lanka has identified "sustainability champions" in its 12 sectors, each of which organizes sustainable skills training in areas such as promoting energy efficiency, waste reduction, and meeting Global Reporting Initiative requirements. Even smaller companies such as Vo Trong Nghia Architects in Viet Nam have developed a niche in sustainable practices by promoting the use of natural and local materials in construction.

Apart from apprenticeships and other traditional in-house training mechanisms, internships have recently gained popularity. By providing internships, enterprises effectively address the deficiency in practical and up-to-date knowledge imparted by TVET institutions. This also proves to be a potential way of recruiting good employees.

Given the large and burgeoning shortage of skilled workers, coupled with poor training and accreditation mechanisms, a number of large corporations, for-profit institutions, voluntary organizations, and NGOs have also started working in the area of human resource development and training. For example, in India, of approximately 10,000 industrial training institutes (ITIs), nearly 7300 are private. Similarly, nearly 73% of TVET institutions in Indonesia are private. There are also efforts by industry associations such as CII and FICCI in India or VBCSD and VEIA in Viet Nam and through the work of international development agencies such as the World Bank, ILO, and ADB to bring in new perspectives and models, flexible funding, high-quality research, and best practices, and to develop individual and institutional capacities to deliver large-scale programs. Although global experience may not provide a solution to all the challenges being faced in these countries; the experience can generate an array of options that could be leveraged and contextualized to meet field-level conditions and realities.

The Green Narrative Is Both Dispersed and Disconnected

The concepts of green workplaces, green jobs, and green skills are relatively new in most countries. There is not as yet a proven set of methods, policies, and approaches to successfully implement green initiatives. While each country has strong environmental policy frameworks that seek to meet global standards, implementation mechanisms at the grassroots continue to be weak. Punitive measures encouraging defaulters to behave differently are lacking, and incentives to urge enterprises to comply are weak. For example, the *Viet Nam National Green Growth Strategy* 2012 seeks to promote green thinking by aiming to reduce GHG emissions and increase clean and renewable energy. Similarly, presidential orders in Indonesia regulate CO_2 emissions, waste management, and renewable energy.

While such initiatives are cutting edge and progressive, in both cases implementation is yet to happen on a large scale.

The actual green practices required in agriculture, industry, and the services sector are quite different. The agriculture sector has connotations of "green" such as water conservation, minimum use of fertilizers, and adoption of appropriate cropping patterns. The services sector is concerned more with conserving energy, while the industry sector is interested in processes such as pollution control, recycling, waste management, procurement, and energy audits. The demand for green skills in traditional industries arises mostly out of the need to comply with regulations. Green skills demands in new industries, such as renewable energy production, arise out of the need to conserve resources and to comply with global sustainability agreements. Within the industry sector, there are different "shades of green" between the sectors that need to be considered when developing strategies.

3 Generic and Specific Green Skills Across Industries

While the demand for green skills in all jobs is being increasingly expressed by industry, not much interest is currently shown in differentiating between green and nongreen jobs. Though academic efforts may continue to identify green jobs across sectors, industry subsectors are more interested in introducing sustainable practices in production and services, which may have implications for creating new job roles that are green or where there is a requirement for top-up skills for the existing workforce.

The evidence is clear that skill shortages are impeding the profitability, productivity, and competitiveness of enterprises. For example, in all four countries, time and cost overruns are one of the major problems faced by the construction industry, these being triggered by a lack of formal training and inadequate systems related to skills assessment and the certification of construction workers. In Viet Nam, the changing preferences of domestic consumers for green buildings are putting pressure on developers and construction companies to develop low-cost as well as sustainable infrastructure options for consumers. Green skills are increasingly required in impact assessments, in using innovative construction materials and building techniques, and for training in the correct use and application of materials and in retrofitting. Some other areas of skills requirements are green certification, green construction techniques, and waste management at construction sites.

Similarly, the labeling requirements of equipment and appliances, efficiency in residential lighting, agricultural pumping, and reducing emission in coal-based energy production are cross-country issues in the energy sector, triggering a demand for specialized skills such as in operating and managing supercritical units, carbon capture and storage, expansion of nonconventional gas resources such as shale gas or coal bed methane, and coal washing. The renewable energy sector, on the other hand, does not identify green skills as an "add-on" in the industry, because it regards the sector itself as green and considers all jobs as green jobs. Yet, skills in project management; installation, commissioning, and grid integration of large-scale renewable energy projects; wind resource assessment; and techno-commercial marketing are increasingly required in wind energy. Similarly, solar energy skills in installing Building Integrated Photovoltaie (BIPV) in buildings; project development skills in handling CSPs; integration of solar PV systems; construction, installation, and commissioning of solar thermal systems; and operation and maintenance skills in troubleshooting of circuitry of solar PV lanterns and home lighting systems are increasingly demanding formal training.

Across all four countries, vehicular emission standards are getting progressively tightened. CNG is being rapidly introduced for public transportation. The changes are demanding skills in different forms for lower as well as higher level operations. For example, rapid mass transport needs top-up skills in driving safety, communications, using complex machinery, and handling hazardous materials. Freight transport includes skills in materials-handling equipment such as stackers, hydraulic or hand pallet trucks, forklift trucks, and jib cranes; and handling hazardous materials; and an ability to handle increased tonnage and higher capacity trucks.

Hospitality (travel and tourism) is one of the largest service industries in all four countries. Although the issue of environmental sustainability is actively being pursued by large and organized corporations, the same is not true for SMEs and in the informal sector. While a number of standards and certifications such as Green Globe, ISO 14001, Earth Check, and Sustainable Tourism Eco-Certification are available for larger corporations, these have not as yet been adopted by smaller enterprises.

Even though an increasing number of enterprises are adopting energy-efficient technologies, specialized people are not as yet in great demand in the hospitality sector. The enterprises are largely dependent on top-up skills for existing staff. Some top-up green skills that the sector is actively seeking include recycling, reusing, and reducing fuel consumption in operations; disposing of non-biodegradable garbage in a responsible way; procuring green materials; converting the use of polythene bags to paper and cloth bags; and using biodegradable cleaning chemicals. Some other emerging roles and green jobs in the sector are homestay owners/workers, golf caddies, polo grooms, nature guides, rural and water/hill adventure guides, security guards, tourist vehicle drivers, and masons for heritage constructions.

While dependence on formal TVET is low, a number of examples from the four countries demonstrate that innovation does not wait for regulation. Innovation is triggered mainly by the intrinsic needs of the enterprises and the preferences of consumers. BIMO Transport in Indonesia commissioned tourist buses that meet European emission standards and immediately became the preferred choice of environmentally conscious tourists from western nations. The Sri Lanka Sustainable Energy Authority (SLSEA) initiated a voluntary program for energy managers and energy auditors, and training is being given for managing air conditioning and efficient combustion in boilers. SLSEA is rapidly expanding, and already has 75 major companies and industries as its clientele.

Some companies such as Colombo Dockyards in Sri Lanka or the Responsible Travel Club of Viet Nam have piloted newer concepts such as recognition of prior learning and remote location tourism involving local communities and have become pioneers in promoting decent work, and providing training that is comprehensive and sustainable. There are many similar examples from India and Indonesia.

As the needs of industries in different sectors rapidly change, the nature of jobs is also changing, becoming more modular than linear in nature. Assembly line production is gradually becoming a thing of the past. Therefore, the training for such jobs also needs to become nonlinear or modular. Multidimensional skills are needed that include an understanding of raw materials or resources, an ability to transform them into final products, along with ensuring sensitivity toward the environment during the whole process. Vocational training or skills development can no longer afford to take a linear approach to the learning of generic content, followed by specific components and later advanced content. Skills development and training clearly need a major overhaul as the global economic and environment scenario changes.

4 Role of TVET in Addressing Industry Needs

The TVET system in all four countries has a colonial past and has evolved with the prevailing legacy of responding to traditional approaches to manufacturing as expected during the Industrial Revolution. While the economy changed, technology prowess dominated, and global trade boundaries lessened, the education and training system was slow to respond and could not match the speed of the changing demands of industry and society. The training system was never able to lead change in the economy and was never responsive to change, being too uncoordinated in approach.

TVET, historically, was a second-choice option in society when compared with more academic, general education. TVET therefore assumed relatively low status over the years and never became as aspirational when compared with many western economies. As a result, the TVET system in all four countries has been grappling with the problems of

- poor relevance of training when compared with the demands of industry;
- unclear pathways for learning, with little convergence with general education or higher education systems;
- low quality of trainers and a lack of professional development of faculty to respond to the changing needs of the economy;
- lack of formal jobs in the organized sector, which contributes to making TVET education unattractive; and
- absence of common standards and certification, and duplication of efforts due to multiple agencies.

The TVET landscape in the four countries needs to be understood by decoupling government and private institutions. While government institutions function to perform a social obligation, private institutions exist to earn a profit, hence the need to respond to market gaps and industry requirements. Any change in TVET has to be triggered through government orders, and decrees for government institutions and industry requirements in the case of private institutions. While government TVET systems are enthusiastic in discussing emerging issues, this should not be confused with sensitivity toward those issues. At an individual level, there may be agreement with changing norms, but at an institutional level there are limited mechanisms in place to encourage or bring about that change.

4.1 Key Impediments to TVET Response to Green Skills

While businesses and industries are slow to respond to adopting sustainable practices, the formal TVET system has yet to come to terms with the rapidly changing economic environment. Less than half of the TVET respondents in Indonesia, Sri Lanka, and Viet Nam agreed that policies and practices are in place that are effective in developing "green" skills to meet industry demands (Fig. 6). However, in India, nearly 60% of respondents agreed to put policies in place for developing green skills. ITIs in India agree that they are able to provide green skills, which is counterintuitive to the overall situation with government ITIs. It could be productive to explore further the reasons why ITIs rate their programs high in green skills. It would also be worthwhile to explore further the kind of challenges that institutions face while addressing the development of green skills (Fig. 6).

While TVET institutions claim to be engaged in green skills development, they also believe demand for green skills from industries is not enough. Except for India, not more than 24% of respondents agreed that there is any demand (Fig. 7). These

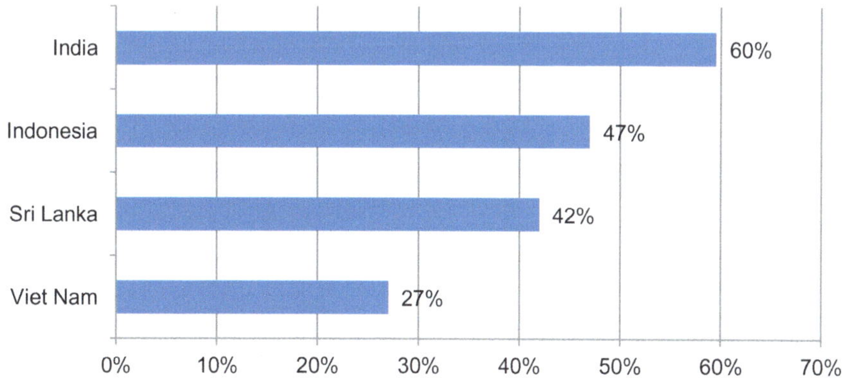

Fig. 6 Effectiveness of Institutional policies and practices in developing students' green skills and knowledge for industry needs. *Source* EdUHK Research Team Survey Questionnaires

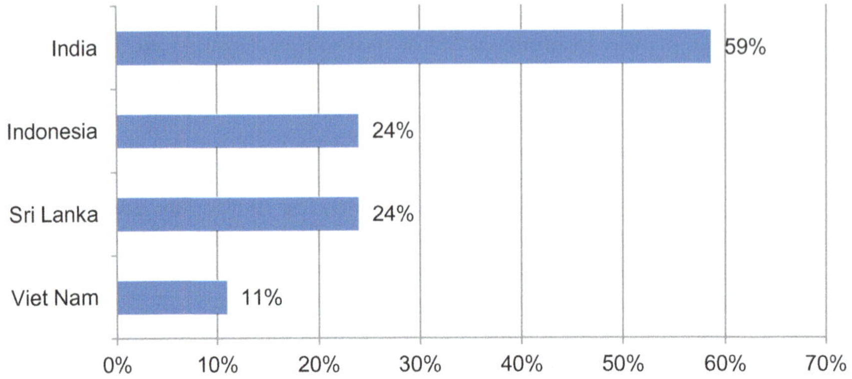

Fig. 7 Demand for green skills. *Source* EdUHK Research Team Survey Questionnaires

findings also confirm that the communication between industry and TVET institutions is passive. The high response rate in India is mainly because of the response of government. Otherwise, NGOs and private sector training providers echo the same message as in Indonesia, Sri Lanka, and Viet Nam.

A further problem in imparting green skills is low individual and institutional capacity. The faculty and trainers of TVET institutions continue to lack updated knowledge and skills in generic as well as green skills. Therefore, the dependence on internal or external partnerships is relatively high. While the dependence is more than 50% in India, it exceeds 80% in Indonesia and is relatively low in Sri Lanka and Viet Nam (Fig. 8). However, the critical need for the professional development of faculty was echoed consistently in both the country interviews and surveys. The high response rate concerning partnerships to impart green skills in Indonesia also emerges from the close linkage of the vocational education system with higher education. Since most training faculty in TVET institutions come from the university system, the presence of vocational education in higher education (as is the case in Indonesia) is highly recommended in other countries as well.

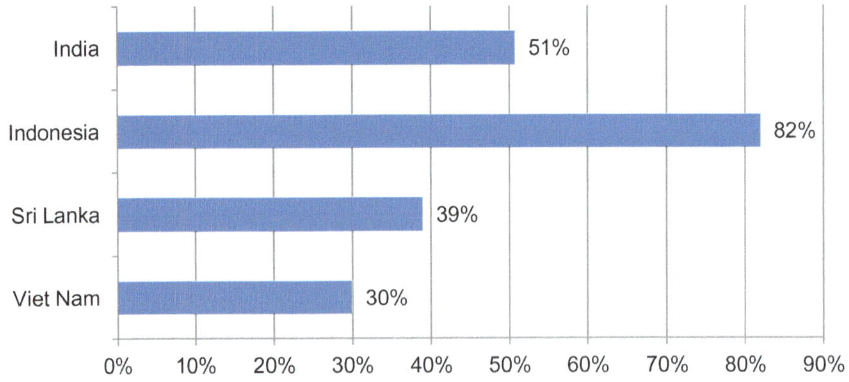

Fig. 8 Dependence on internal or external partnerships to provide green skills training. *Source* EdUHK Research Team Survey Questionnaires

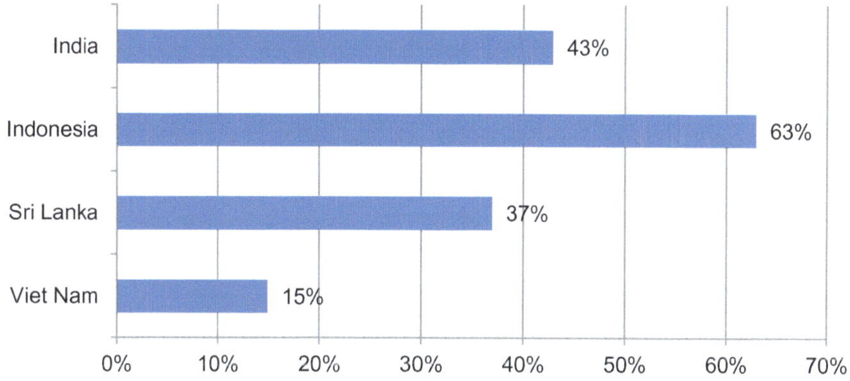

Fig. 9 Green Content learned on the job as part of training programs. *Source* EdUHK Research Team Survey Questionnaires

Poor institutional capacity also leads to higher dependence on industry exposure visits to learn green content. While more than 60% of TVET institutions in Indonesia reported industry exposure as being critical to learning green content, more than 40% in India, 37% in Sri Lanka, and only 15% in Viet Nam agreed to the importance of industry exposure to learn green skills (Fig. 9). It is likely that, among those who believe otherwise, their understanding is limited to providing theoretical inputs as part of general training content, such as environmental awareness subjects, conservation, or recycling.

Another trigger for the lack of faculty development, and the overall passive approach of TVET institutions toward green skills development, is the absence of green skills standards or certification systems. Apart from Indonesia, where 56% noted having green standards, the other countries reported on the poor availability of any standards related to green skills (Fig. 10). Similarly, the availability of certification on green practices was reported as being lowest in Sri Lanka and Viet Nam and just above 35% in India and Sri Lanka (Fig. 11).

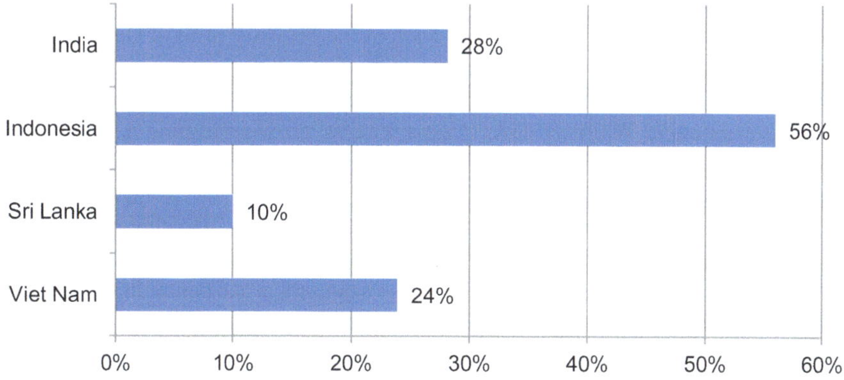

Fig. 10 Existence of Green skills standards. *Source* EdUHK Research Team Survey Questionnaires

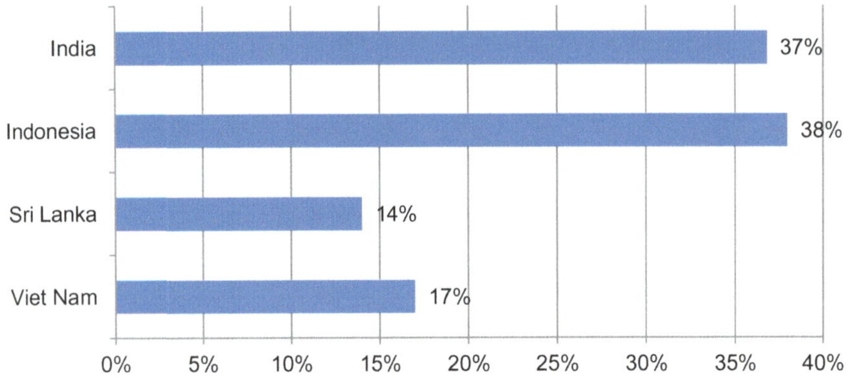

Fig. 11 Availability of green skills certification. *Source* EdUHK Research Team Survey Questionnaires

4.2 *What Works? Initiatives from Countries to Fast-Track Reforms in TVET*

Evidence shows that skill shortages may be impeding the transition to green growth. However, the demand from industry sectors for jobs that are environmentally sustainable and decent is on the rise. Initiatives within industries and the society as a whole are needed to develop awareness about sustainable practices and green employment. TVET cannot work in isolation, but must be aligned with economic policies and with new climate change and industry requirements. Standard approaches to addressing issues within education and training urgently need to be revisited and aligned with sustainable growth. Although the problems appear to be enormous, the range of innovative and appropriate activities through the efforts of government, industry, private entrepreneurs, NGOs, and international institutions, should not be ignored.

For example, one of the key strategies adopted by the Government of India has been to put skills development as a high priority on political agenda. The government has set up dedicated institutions such as the Ministry of Skill Development, Entrepreneurship, Youth Affairs and Sports with dedicated agencies such as NSDC and NSDA, to channel and facilitate private sector and government sector efforts in a cohesive manner. While NSDA is encouraging cross-institution learning and sharing, NSDC is extending loans to the private sector for skills development training. Initiatives by other ministries, such as *Hunar Se Rozgar* (skill to employment) of the Ministry of Tourism or "*Aajeevika* Skills" (livelihood skills) of the Ministry of Rural Development, are being implemented through NGOs. The private sector is also actively engaging in developing a sustainability-orientated curriculum, green skill courses, and certification systems. Similar initiatives such as "Greening TVET" in Viet Nam, "Factory in Schools" in Indonesia, or the "*Harita Lanka* Programme" in Sri Lanka are important policy directions and platforms to fast-track reforms in the TVET system.

All four counties demonstrate models of TVET practices that are at different stages of development but that establish some underlying principles critical to initiating general skills development or green skills development reforms.

Initiatives Driven and Sustained by Apex Government Agencies, Large Corporations, and Strong Leadership are Most Likely to be Successful

NSDC in India is a unique initiative between the government and the private sector to expand the reach of skills development through private training providers, as well as to instill quality by setting quality standards in training, and occupational standards through sector skill councils (SSCs). This initiative is backed by strong government funding and corporate sector support of the highest order. Two leading industry corporations, CII and FICCI, are key stakeholders in NSDC. In addition, leading sector-specific industries are part of the SSCs initiated by NSDC. It is also important to note that the leadership of NSDC has not changed since its formation in 2009.

In a similar way, the National Cleaner Production Centre (NCPC) in Sri Lanka has emerged as the foremost cleaner production solutions provider there. NCPC was established by UNIDO under MOIC to help enterprises adopt cleaner production. NCPC has also expanded due to continuous government and multilateral agency support.

Initiatives Catering to International Demands to Meet International Standards and Global Mandates are Easily Acceptable

The Brandix green factory in Sri Lanka has surpassed all green factory standards stipulated for energy consumption, water conservation, solid waste management, and carbon emissions by the Leadership in Energy and Environmental Design Green Building Rating System of the US Green Building Council. It is one of the leading plants manufacturing apparel for Marks and Spencer. The Brandix Eco Centre provides on-the-job training to all employees in clean technology, waste management and environmental preservation, ongoing environmental awareness raising, wastewater management, and solid waste disposal through reuse or recycling. Other apparel enterprises such as the Hirdaramani Group in Sri Lanka have also achieved high levels of international standards and made Sri Lanka into the only outsourced apparel manufacturing country in Asia that has signed up to 30 ILO conventions. It stands out as a reliable sector that pays fair wages to its workers while discouraging sweatshops and child labor in its supply chain businesses. Environmental issues are also given precedence.

Similarly Balai Latihan Pendidikan Teknik (BLPT) in Indonesia, which was established in association with a provincial government, is focused on developing the quality of TVET graduates to meet international standards. BLTP offers its services not only to its own students but also to other TVET students in the region who do not have the opportunity to gain practical exposure. BLTP has a production center that meets international standards for waste management and recycling, and it has a number of industry partners in car parts production, tractor manufacturing, and electric tricycle production. BLTP sends its instructors to the Vocational Education Development Centre for professional development and for benchmarking with other centers. PT Esemka is another leading example, where students make cars that are sold in the market. The cars are built adopting international standards and are competitive with leading car manufacturers in the world.

Initiatives Having Large Social Demand are Sustainable

The Tran Hung Dao Vocational College in Viet Nam trains farmers in clean production to improve their productivity and develop the ability to export clean products. The short-term training courses provided under their rural training policy strategy bring in experts from the Vietnam National University of Agriculture and the Bristol Research Institute of Agriculture to consult on curriculum development and technology. The college is in great demand among rural youth and farmers in Viet Nam.

Likewise, Basix Academy for Building Lifelong Employability, an NGO in India, runs programs on green skills in the province of Sikkim, the only organic agricultural production state in India. In the design of its courses, it uses a green skills inventory (which identifies training classes) that incorporates agriculture, forestry, animal husbandry, fisheries, and horticulture. Some classes are in green construction, retrofitting for energy efficiency, green personal and social services, and basics of green habitats.

Both programs are able to sustain and expand their work because of the popularity and utility of their programs for their catchment communities. Such programs survive and become successful because of a high degree of local support and very low dependence on external funds and resources.

Initiatives Should Have Sound Processes Such as Strong Communication, Monitoring, Reporting, and Should be Easily Replicated

There are many examples from the four countries studied wherein institutions that have established strong operational management systems and standard operating procedures are popular, and their work easily becomes replicable. NSDC in India has supported nearly 31 SSCs to develop accreditation systems as well as occupational standards for industries. Replication has been possible mainly due to strong standard operating procedures across the SSCs. NSDC is now keen to promote green occupational standards through the SSCs. Similar rigor is evident in MSME tool rooms and CIPET institutions in India.

Don Bosco in Sri Lanka aims to develop a coherent approach toward greening, including the development of a green campus across its centers throughout the country. They have developed common guidelines and documentation on waste management, the use of biogas for cooking, and the recycling of scrap metal and timber. Likewise in Indonesia, eight TVET schools on Java Island have created a framework on sustainable practices.

Each of the examples presented demonstrates the power of the principles of leadership, international acceptance, social demand, and standardization toward triggering change, be it in reforming the TVET system or introducing a green skills mandate. To be successful, policy needs to make way for the adoption of such principles, the implementing cadre needs to build its capacity to carry forward the agenda, and consumers need to be sensitive and responsive to the intended transformation.

5 Potential Areas for Further Research

This research has been an exploratory study of education and skills for inclusive growth and green jobs, and the greening of the economy in India, Indonesia, Sri Lanka, and Viet Nam.

The research has highlighted the need to undertake further research in the following areas in order to fill the gaps that exist in the current data, with regard to:

(i) **Examining various geographical parts of each country**. The respondents to the survey questionnaires and the interviewees were mainly from the capital cities and nearby areas in each of the four countries. India, Indonesia, Sri Lanka, and Viet Nam are diverse countries in terms of rural as well as urban characteristics, and with regard to the socioeconomic status of their different regions. It would therefore be useful to undertake the same survey and conduct interviews with population groups in other parts of each of these four countries.

(ii) **The importance of the informal and nonformal economy, sometimes called the disorganized sector**. This study focuses on the formal, organized economy, but much of the economic activity in the four countries studied occurs in the disorganized sector. It would be helpful therefore to also survey the nonformal economy to ascertain whether the findings from this study also hold true for it.

(iii) **The rural sector**. The emphasis in this study has been on various secondary industries such as construction, transport, and hospitality. But most employment in India, Indonesia, Sri Lanka, and Viet Nam currently occurs in the rural sector, so it would be relevant to undertake a study to examine the same issues, concerns, and challenges among those working in the rural sector.

(iv) **Small and medium-sized enterprises**. SMEs are underrepresented in this study, yet most businesses in the four countries are SMEs. It would be helpful to focus more on SMEs by administering to them the same survey instruments and conducting interviews.

(v) **Gender issues with regard to inclusive growth and green jobs**. This research study did seek to gather information about and survey the matter of gender issues, but overall there is a paucity of information available and a lack of reliable research evidence in this area, most the data being anecdotal in nature.

(vi) **Unemployed and underemployed youth**. Likewise, in the case of youth, with particular reference to unemployment and underemployment, there is a lack of useful or reliable evidence. Given the major problem of youth unemployment, further research in this area would be helpful for policy makers.

In addition, this ADB study has examined four countries in Asia: India, Indonesia, Sri Lanka, and Viet Nam. There is rich potential to study other important countries in the Asia and Pacific region using the same research methodology, survey instruments, and interview schedules as have been used for the four countries reported on here. It could, for example, be helpful to study the Republic of Korea, the People's Republic of China, Thailand, and Malaysia, and also some of the Pacific Island economies, for comparison purposes. The People's Republic of China would be a particularly interesting country to study, because, as a country that faces major environmental problems, it is at the forefront, internationally, and very proactive in promoting green jobs, the greening of the economy, and inclusive growth through innovative means and stressing the replication of best practice.

If further country studies were undertaken, for comparison purposes these could be based on the same surveys and interview research instruments as used in this study, but in a modified form to take account of any of the shortcomings of the current study, particularly regarding areas of coverage as referred to above, and to take account of the varying contexts of the countries concerned regarding any additional studies undertaken.

6 Concluding Comments

The four country studies of India, Indonesia, Sri Lanka, and Viet Nam, and the regional analysis, undertaken as part of *ADB Project TA-7879 REG: Education and Skills for Inclusive Growth and Green Jobs in Asia* is unique in this field. Unlike other studies undertaken in Asia into this area, which rely and report on mainly secondary sources of information, this study has relied largely on primary data collected through surveys and interviews with key players including those in government, business enterprises, business associations, TVET providers, international development agencies, NGOs, and others in each of the four countries studied. These primary data, which are unique in this field for Asia, are the main value added of this research study.

Skills development and training need to be overhauled as the global economic and environmental scenario changes in Asia. Vocational training or skills development can no longer afford to take a linear approach to learning generic content, followed by specific components and later by advanced content. What is required is a nonlinear approach, because jobs today are becoming more modular in nature. Multidimensional skills are needed that include the understanding of raw materials and resources, and the processes that transform them into final products, with sensitivity toward the environmental issues embedded into the whole process. Therefore, generic green skills are viewed by industry as measures to reduce costs and increase profitability, suggesting that the holders of these skills are valued by employers.

The labor market across industry sectors is demanding jobs that are more environmentally sustainable and decent, as there is growing evidence that skills shortages may be impeding the transition to green growth in some sectors. Taking a wider view of green skills, similar to the ILO's decent work agenda, initiatives should be integrated into local communities and industry that can produce results leading to new business opportunities and further generate supplementary green employment.

Policy makers and practitioners have a major task to balance emerging needs and priorities with existing implementation frameworks. The balance has to accommodate norms and standards from global agreements, whether they be the Millennium Development Goals or climate change agreements. They need to tread a fine line between industry demand, economic priorities, and environmental mandates or inclusive development.

Fortunately, there is no lack of knowledge about the subject. Knowledge from within and about other countries, and partnerships with international agencies, are available to bring new perspectives; new models; innovative practices; flexible funding; and high-quality research to develop individual and institutional capacities to deliver large-scale programs around green economy, green jobs, and green skills.

Appendix A
List of Generic Green Skills

Skills area	High skilled	Medium skilled	Low skilled	Civil society
	The design, analysis, and evaluation of systems/ processes for products and services	The maintenance and monitoring of systems and processes for products and services	The implementation of processes through the production of products and services and recording of information	The knowledge and skills to make responsible choices and maintain a sustainable lifestyle
Environmental awareness				
What is global warming and what is the environment What are natural resources; what is climate change; and why is it occurring What is the impact of climate change and what can be done about it	√	√	√	√
What are the responsibilities of society and industry What are water conservation, energy efficiency, waste segregation, recycling	√	√	√	√

(continued)

R. Maclean et al., *Education and Skills for Inclusive Growth, Green Jobs and the Greening of Economies in Asia*, Technical and Vocational Education and Training: Issues, Concerns and Prospects 27, https://doi.org/10.1007/978-981-10-6559-0

(continued)

Skills area	High skilled	Medium skilled	Low skilled	Civil society
What is happening within the industry sector at a national and international level	√	√		
What are the drivers and benefits	√	√	√	
What are best practice examples of sustainable business and what are the cost savings	√	√	√	
What are the changes in business processes to become more sustainable	√	√		
How do you introduce new sustainable systems	√			
Energy efficiency				
Why energy efficiency is important Energy-efficient technology and energy-efficient labeling: what they mean What can individuals do to be energy-efficient in the home and at work	√	√	√	√
Undertaking a basic energy audit	√	√	√	√

(continued)

(continued)

Skills area	High skilled	Medium skilled	Low skilled	Civil society
What sector-specific procedures best support energy efficiency	√	√	√	
Design energy-efficient production procedures, evaluate, and improve	√			
Monitor approaches in energy efficiency in production processes		√	√	
Calculating the costs of energy and the savings of energy efficiency	√			
Comparing costs of sector-specific energy-efficient technology	√			
Calculating cost savings from energy efficiency measures and return on investment sector wise	√			
Water conservation				
Why water conservation is important What is water efficiency Water-efficient devices and what can individuals do to be water efficient in the home and at work	√	√	√	√

(continued)

(continued)

Skills area	High skilled	Medium skilled	Low skilled	Civil society
Undertaking a basic water-efficiency audit	√	√	√	√
What sector-specific procedures best support water efficiency and conservation	√	√	√	
Comparing costs of sector-specific water conservation and waterless technology	√			
What water conservation systems work best in different environments	√			
Design water conservation production procedures, evaluate and improve	√			
Monitoring approaches in water conservation in production processes		√	√	
Calculate the costs of water conservation and the savings	√			
Calculate cost savings from water conservation measures and return on investment sectorwise	√			

(continued)

(continued)

Skills area	High skilled	Medium skilled	Low skilled	Civil society
Waste reduction and waste management				
Why waste reduction and recycling are important What are waste reduction and recycling Waste segregation and recycling and what can individuals do in the home and at work	√	√	√	√
Approaches to waste management, segregation, waste recycling, and waste reduction for solid and liquid waste	√	√		
What sector-specific procedures best support waste reduction and waste management	√			
Comparing costs of sector-specific waste reduction, waste recycling, and waste management	√			
What waste recycling and waste management systems work best in different environments	√			

(continued)

(continued)

Skills area	High skilled	Medium skilled	Low skilled	Civil society
Design waste reduction and waste recycling production procedures, evaluate and improve	√			
Monitoring approaches in waste reduction and waste recycling in the production processes		√	√	
Calculate the costs of waste reduction and the savings	√			
Calculate cost savings from waste reduction, waste recycling, and waste management measures and return on investment sector wise	√			
Auditing and evaluation				
Certification/ audit/ verification process skills		√	√	
Certification/ audit/ verification process management skills	√	√		
Environmental data analysis	√	√		
Auditing	√	√		
Evaluation	√			

(continued)

(continued)

Skills area	High skilled	Medium skilled	Low skilled	Civil society
Knowledge and understanding of international and national standards and legislation				
Environmental policies	√	√		
Compliance	√	√	√	
Mapping against standards	√	√	√	
Understanding systems				
The environment as a system • Society as a system	√	√	√	√
The dynamics, components, and relationships of different systems	√	√	√	√
The production process as a system	√	√	√	
Designing systems; identifying the processes to transform inputs into outputs	√	√		
Identify problems as part of an overall system	√	√	√	√
Identify improvements as part of an overall system	√	√	√	√
Identifying new connections	√	√	√	√
Examine cyclical cause and effects across the system	√	√	√	√

(continued)

(continued)

Skills area	High skilled	Medium skilled	Low skilled	Civil society
Evaluate the linkages and interactions between the elements that compose the entirety of the system	√	√	√	√
Collaboration skills				
Collaborative thinking, identify the benefits and sustainability opportunities for collaboration	√	√	√	√
Teamwork, compromising, and communication skills for collaboration	√	√	√	√
Develop relationships and work with others to achieve common sustainability goals and a shared purpose	√	√	√	√
Identify opportunities to mutually benefit from shared resources and skills	√	√	√	√
Develop behaviors to problem solve and achieve the objective of the collaboration	√	√	√	√
Identify collaborative opportunities to progress sustainable business goals	√	√		

(continued)

(continued)

Skills area	High skilled	Medium skilled	Low skilled	Civil society
Innovation skills				
Innovation thinking, creative thinking, continuous improvement, risk taking, collaboration	√	√	√	√
Identify, and map needs or opportunities	√	√	√	√
Identify and map needs or opportunities related to your workplace	√	√	√	
Identify and map needs or opportunities related to your organization's sustainability goals	√	√		
Identify needs or opportunities and map using a system thinking approach	√	√	√	√
Generate a number of ideas or solutions	√	√	√	√
Generate a number of ideas or solutions related to your workplace	√	√	√	
Generate a number of ideas or solutions related to your organization's sustainability goals	√	√		
Implement ideas or solutions	√	√	√	√

(continued)

(continued)

Skills area	High skilled	Medium skilled	Low skilled	Civil society
Implement ideas or solutions related to your workplace	√	√	√	
Implement ideas or solutions related to your organization's sustainability goals	√	√		
Evaluate a number of ideas or solutions	√	√	√	√
Evaluate a number of ideas or solutions related to your workplace	√	√	√	
Evaluate a number of ideas or solutions related to your organization's sustainability goals	√	√		

Generic Green Skills for Medium to Large Organizations

Skills area	High skilled	Medium skilled	Low skilled
	The design, analysis, and evaluation of systems/ processes	The maintenance and monitoring of systems and processes	The implementation of processes through the production of products and services and recording of information
Auditing and evaluation			
Certification/audit/ verification process skills		√	√
Certification/audit/ verification process management skills	√	√	
Environmental data analysis	√	√	
Auditing	√	√	
Evaluation	√		

(continued)

(continued)

Skills area	High skilled	Medium skilled	Low skilled
Knowledge and understanding of international and national standards and legislation			
Environmental policies	√	√	
Compliance	√	√	√
Mapping against standards	√	√	√
Stakeholder engagement			
Collaboration	√	√	
Stakeholder management	√	√	
Product life cycle impact assessment			
Impact measurement/ assessment	√	√	
Compilation of inventories of relevant energy and material inputs and environmental releases		√	√
Evaluation of the potential impact of identified inputs and releases	√		
interpreting results to inform decision-making	√	√	
Green procurement	√	√	
Identify and order green resources, products, and services	√	√	
Determine ratings and test results of resources and products	√	√	
Evaluate the potential impact of using one product, resource or service over another on the overall end product or service	√		
Interpreting results to inform decision-making	√	√	

Source EdUHK team analysis

Appendix B
Project Background and Introduction, Approach, and Methodology

The scope of the Asian Development Bank (ADB) commissioned research project "Education and Skills for Inclusive Growth and Green Jobs" was to increase the knowledge and capacity of the four developing member countries (DMCs) being studied (India, Indonesia, Sri Lanka, and Viet Nam) to match education and skills training to jobs, including green occupations. This research study seeks to examine how the four DMCs can best put in place timely and effective policies and strategies for skills development. It also aims to facilitate dialogue and networking among the public sector, business and sector councils, industry and occupational associations, and employee associations to assess gaps in policy and practice for the development of skills and to anticipate future needs.

B.1 Overall Project Methodology

Green growth requires new green sectors and activities to be developed and new skills for both new jobs and existing jobs that are changing in response to carbon constraints. Therefore, skills development policies can make an important contribution to green growth. The research study reported here attempts to provide a link between green growth and education and training provision. International Labour Organization (ILO) and European Union (EU) (2011a, b) reports also suggest that research focusing on economic-related measures and indicators of green growth is required for green growth, in conjunction with research into skills.

While some studies aim to improve transparency in the training markets in particular sectors and some aim to define the content for new training programs in terms of skills and topics based on employer surveys, others have more complex approaches toward determining skills needs, training and employment policies through sectoral committees that represent different stakeholders (ILO and EU 2011a, b). This study does not focus on skills characteristics as such, but on the training that is available for green skills and on the ways they could best be introduced into technical and vocational education and training (TVET) curricula.

One of the unique features of this study is the primary surveys conducted with a significant number of TVET institutions and industries in selected sectors in

R. Maclean et al., *Education and Skills for Inclusive Growth, Green Jobs and the Greening of Economies in Asia*, Technical and Vocational Education and Training: Issues, Concerns and Prospects 27, https://doi.org/10.1007/978-981-10-6559-0

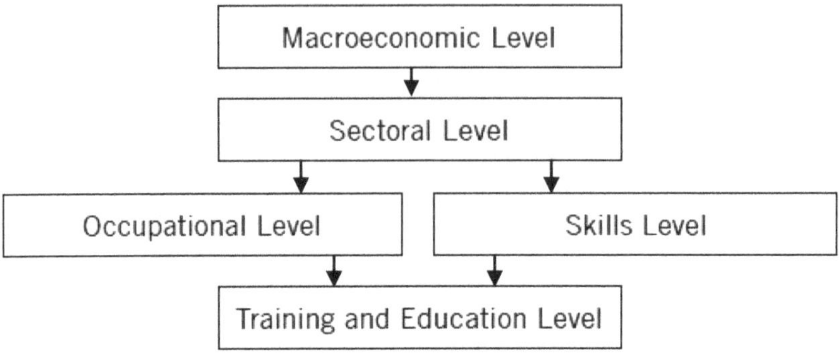

Fig. B.1 Levels of analysis in skills research. *Source* ILO and EU 2011a, b, p. 5

addition to personal interviews and focus group discussions with important stakeholders. The evidence was collected through quantitative surveys, which were later validated through interviews and focus group discussions. Furthermore, expert opinion was also sought through workshops and roundtable discussions involving all the stakeholders in order to authenticate the findings and the subsequent policy advice.

Key stakeholders were identified and consulted both from the labor skills demand side [high-growth selected industries, namely, manufacturing, energy, construction, transportation, and services (tourism and/or hospitality)] and from the supply side [formal government TVET providers and informal private and/or nongovernment organization (NGO) skills trainers] with a special emphasis on emerging green skills and green jobs. It was important to evaluate the extent to which training providers respond to the needs of industry and, second, to the needs of green growth. The project was therefore designed across sectoral, occupational, skills, and training levels (Fig. B.1).

Challenges that need to be met for any research include the need to develop a research approach that is reliable, transparent, and appropriate to the context of specific countries. The following section provides details about the approaches adopted for different components of the study.

B.2 Methodology for Industry and Enterprises

Relevant data were collected from various companies through surveys and interviews. The survey and interview questionnaires were prepared after reading the research on green jobs, sustainable company practices and by focusing on the study objectives and common issues that needed to be addressed including (but not limited to) a lack of awareness on the part of many enterprises in the sector concerning greening issues and their importance, and problems associated with a shortage of skilled labor that has green skills. Interview questions supplemented the survey by addressing questions concerning the higher costs associated with the greening of processes and the importance of government policy, legislation and incentives to encourage the greening of production and services; and the value of setting industry standards and industry self-regulation to promote the greening of enterprises.

The research employed a multimethod approach involving a desktop review of the literature and relevant grey documents. The in-country component of this study consisted of 2 weeks of interviews in each country involving senior members of enterprises and senior representatives from organizations comprising industry associations, NGOs, government, and international donor organizations.

A survey was developed based on extensive reading of the literature related to green jobs, cleaner production, and engaging the business industry in sustainability. A separate survey of TVET training providers was also undertaken. A roundtable of stakeholders was also held.

Due to limitations in the size of this study and the geographic spread of the interviews, which were in the capital cities of the countries in the case studies, additional secondary sources and government planning documents were also considered as they provide the national context for determining findings that are relevant nationwide. Some secondary industry research data have been reanalyzed through the lens of green skills to further add to this study.

The responses were sorted and analyzed to establish findings indicating the skills gaps, issues, and challenges that have important implications for government, the private sector, and NGOs. The employer interviews and surveys undertaken as part of this study have not only validated earlier research but have resulted in a clearer understanding about what generic green skills are and how they might be applied. These are elaborated upon in this report and are reflected in the policy recommendations arising from this study.

B.3 Methodology for TVET Providers

Qualitative questionnaire surveys and interviews were used to elicit informed opinion and expert knowledge on structured information relevant to current training practices and issues, as well as green skills. This was undertaken in a structured way that helped to capture the current practices of TVET providers in terms of approaches used by institutions to respond to skills needs; to government policies and initiatives that are relevant to skills development; and to economic restructuring and staff training. A particular emphasis was placed on approaches used by institutions to include green skills (skills that are required to reduce the adverse environmental impact of enterprises and the economic sectors, ultimately to levels that are sustainable) in their programs and courses. Representatives from government, NGOs, and donor agencies were also interviewed.

B.4 Methodology for Identifying Skills Profiles

The industry survey data and interviews were the primary source for identifying green skills and the evolving nature of jobs in green industry sectors associated with this study. Additional information was sourced from job advertisements, company career information, and relevant green industry associations. The resulting information was also validated against internationally available green or sustainable national occupation standards (competency or skills standards), documented sustainable businesses practices, and cleaner production processes.

B.5 Outcomes of the Report

This study found that the skills profiles of workforces that encourage successful sustainable growth are more likely to involve the widespread use of generic green skills. Furthermore, specific underpinning knowledge of sustainable technology or processes and greater awareness of environmental issues related to work practices combined with these generic green skills vary in complexity depending on how they are applied. According to industry, these generic green skills lead to cost reductions as well as reducing the individual company's carbon footprint. The identification of generic green skills will allow governments and industry to develop strategies for mainstreaming these skills across the labor market and potentially reducing carbon footprints of individual enterprises and value chains. Training and education providers will be able to integrate these generic green skills into existing training programs to improve the sustainable work practices of TVET graduates.

The study identified a number of generic green skills and that many of the new green jobs draw on hybrid skills sets; for example, installing solar hot water requires low-level electrical and plumbing skills. The interviews highlighted that most of the generic green skills have elements that occur at different levels of complexity and operational independence.

The contents of this report can be put to use by skills development NGOs, NGO partners such as public sector undertakings, corporate and noncorporate donors who fund skills development programs, and policymakers alike. The green practices identified for skills development programs can also help align TVET with the larger picture of green jobs for a greener and more sustainable economy.

Appendix C
Employers' Survey Instrument

Please read each question carefully and tick whichever response is appropriate

Please add your names and contact details if you are happy to be interviewed.

We value your views

Questionnaires will be analyzed by the project researchers and will be treated as strictly confidential.

Background Information on the Project

This survey is for an Asian Development Bank project on Education and Skills for Growth and Green Jobs. The information that you provide will help us identify successful practices in addressing industry needs in skills development, challenges, and requirements for training providers in the Technical Vocational Education and Training (TVET) sector to develop training programs with specific focus on meeting the changing skills needs of the labor market due to demands of climate change and environmental sustainability issues. Most of the questions require you simply to tick a box but there is an opportunity at the end of the survey to add any details you feel are important, but have been missed in earlier questions.

The Asia-Pacific region uses three times the resources as the rest of the world to create one unit of GDP, and between 2000 and 2005 the resource intensity of its economy increased, in contrast to trends in the rest of the world (*UN* 2012a, b).

Green jobs are jobs that reduce the environmental impact of enterprises and economic sectors, ultimately to levels that are sustainable (ILO.)

Greening is a term for moving from traditional production processes or services to production processes or services that have a reduced environmental impact.

The Asia-Pacific region, while emitting relatively low levels of greenhouse gases on a per-capita basis, is one of the fastest-growing sources of climate impacting emissions (UN 2012a, b). *This is the context in which this survey is*

R. Maclean et al., *Education and Skills for Inclusive Growth, Green Jobs and the Greening of Economies in Asia*, Technical and Vocational Education and Training: Issues, Concerns and Prospects 27, https://doi.org/10.1007/978-981-10-6559-0

being conducted. Comparative surveys are being held with a sample of members of government, nongovernment, and private sector business and industry organisations.

Thank you,
Prof. Rupert Maclean
Chair Professor of International Education
The Education University of Hong Kong

1. **Does your business USE any of the following?**

	Yes	No	Would like to but do not have worker skills	Would like to but do not have the technology	Would like to but do not have business processes
Recycled materials					
Energy-efficient products/services					
Energy-efficient infrastructure like solar power, water recycling, waste disposal, etc.					

2. **Does your business PROVIDE any of the following?**

	Yes	No	Would like to but do not have worker skills	Would like to but do not have the technology	Would like to but do not have business processes
Recycled materials to others					
Energy efficiency to others through your products/services					
Energy-efficient infrastructure to others like solar power, wind power, water recycling, waste disposal, etc.					

3. **How important are the following to your company for becoming environmentally friendly**:

	Very Important	Important	Neutral	Not so Important	Not Important
Innovation in technology, products, processes, and services?					
Collaboration with others especially with those in the supply/value chain?					
Requirement or need to undertake impact assessments related to climate change or environmental issues?					
Understanding environmental standards and legislation?					
Technical and Vocational Training (TVET) of employees?					

4. **How much importance you give to government policies on**

	Very important	Important	Neutral	Not so important	Not important
Economic development policies?					
Environmental policies?					
Skills development policies?					
Green job policies?					

5. **To what extent the following issues have affected your business in the last 12 months?**

	Greatly	To some extent	Cannot say	Not much	Not at all
Government legislation, regulations or requirements					
Industry standards and competition					
Rising consumer demand for green products/services					
Increasing costs (e.g., cost of energy)					

(continued)

(continued)

	Greatly	To some extent	Cannot say	Not much	Not at all
Ethical/corporate responsibility issues					
Demand for "green skills" of workers					

6. **To what extent have the following forced you to develop new skills in your company?**

	Greatly	To some extent	Cannot Say	Not much	Not at all
The introduction of new international standards (ISO, OHS related to climate change or environmental issues)					
Corporate social responsibility related to climate change or environmental issues					
Import–Export requirements related to climate change or environmental issues					
Government legislation related to climate change or environmental issues					
Requirement or need to undertake impact assessments related to climate change or environmental issues					
Customer/ client expectations related to climate change or environmental issues					

7. **In the last 12 months, has your company….**

	Yes	No	Can't Say
Needed to employ new workers or train employees in new skills related to dealing with climate change or environmental issues?			
Needed to integrate jobs related to dealing with climate change or environmental issues and inclusive growth into work environment?			
Witnessed growing awareness amongst senior managers on climate change or environmental related jobs?			
Faced a shortage of skilled workers in climate change or environmental related jobs which might be detrimental to the business?			
Experienced opening up of new markets or business lines related to climate change and environment?			
Needed to account for inclusive policies in labor force planning?			

8. **Do you want your business to become more environmentally friendly in the following areas?**

	Yes	No	Would like to but do not have worker skills	Would like to but do not have the technology	Would like to but do not have business processes
Involvement in recycling					
Involvement in new green products or services					
Energy-efficient infrastructure like solar power, water recycling, waste disposal, etc.					
Become more informed on green issues					

9. **What is the possible impact on your business in the absence of any jobs related to climate change?**

- ☐ We are not able to expand our work (business)
- ☐ We are not working as well as we could be
- ☐ Our costs are higher than they should be
- ☐ We have higher workloads on our existing workers
- ☐ We have problems maintaining our equipment
- ☐ We cannot meet our sustainability target goals
- ☐ No effect

10. **Does your business regularly review its employee skills with respect to changing business requirements?**

- ☐ Yes on its own initiative
- ☐ Yes, at the request of outside bodies
- ☐ Cannot say
- ☐ No

11. **Does your business use any of the following data sources or forecasting tools/mechanisms to make decisions about what employee skills will be needed? (Please tick)**

- ☐ Business research (including market surveys)
- ☐ Published statistical data
- ☐ Government directions/advice
- ☐ Labor market analysis
- ☐ Industry magazines
- ☐ Job announcements analysis
- ☐ Links with TVET institutions

12. **Methods of training used in your organization? (Please tick all that is applicable)**

☐ Professional development like time management, decision-making, planning, etc.
☐ Mentoring new joinees/ Buddy system
☐ Internal technical training on operations
☐ Training of your employees by vendors or distribution channels
☐ Apprenticeships/ Internships/ Industrial placement—students placed in your company
☐ Sponsoring employees for external training programs
☐ Capacity building through attending conferences, trade shows, industry association meetings, union meetings, etc.

13. **With respect to your suppliers how important are the following to your business**

	Very important	Important	Neutral	Not so important	Not important
Suppliers using skilled and trained workers					
Suppliers adhering to international standards in manufacturing					
Suppliers engaged in recycling of materials					
Suppliers producing energy-efficient products					
Suppliers using energy-efficient infrastructure like solar power, water recycling, waste disposal, etc.					

14. **How do the various institutions mentioned below respond to your company's skill requirement?**

	Greatly	To some extent	Cannot say	Not much	Not at all
Polytechnics					
ITIs					
Vocational education in secondary schools					
Private professional colleges					
Universities					
Other					

15. **As part of work force planning do you have any involvement with TVET institutions in providing feedback on student/graduate skills and providing input on changes to courses/programs?**

- ☐ Yes, frequently
- ☐ Yes, occasionally
- ☐ Cannot say
- ☐ Not at all

16. **In what ways, if any, does your business engage with TVET institutions?**

- ☐ Provide workplace for on-the-job training (industrial placements)
- ☐ Member or Chair of accreditation panels at TVET institutions
- ☐ Engaged in content development of TVET courses
- ☐ Allow site visits for students as part of their course
- ☐ Provide TVET teachers with industry experience

17. **Would you like to have more opportunities to engage with TVET institutions in various ways?**

- ☐ Yes, frequently
- ☐ Yes, occasionally
- ☐ Cannot say
- ☐ Not at all

18. **What are the key changes related to greening in your business/sector? Please tick as many as are appropriate**

- ☐ Job role changes (including knowledge and basic skills, specialized and technical skills, attitudes, and other personal traits)
- ☐ Changes of enterprise systems (processes and procedures)
- ☐ Changes in work systems (including people, technology, infrastructure, work environment, and materials)
- ☐ Changes in value chain requirements and processes

19. **Your Name/Designation/Mobile/Email**

20. **Name of Company**

21. **What is the Industry sector to which your business/company belongs to?**

- ☐ Construction
- ☐ Automobile
- ☐ Hospitality
- ☐ Energy
- ☐ Others

22. **Type of ownership of your company**

- ☐ Private/Family Owned
- ☐ Multinational

- ☐ Govt. owned
- ☐ Autonomous under Govt.

23. **What is the location of your business where bulk of production or services happen?**

- ☐ Metropolitan
- ☐ Regional (Urban)
- ☐ Regional (Rural)

24. **Would you like your company name to be acknowledged in the published report?**

- ☐ Yes
- ☐ No

Appendix D
TVET providers' Survey Instrument

Asian Development Bank (ADB) along with The Education University of Hong Kong (EdUHK) has undertaken an ambitious project across four nations Viet Nam, Indonesia, Sri Lanka, and India to increase knowledge and capacity of the member countries to put in place timely and effective policies and strategies toward developing skills for environmentally sustainable jobs or green jobs. The project will facilitate dialogue between the public sector, business sector, academic councils, and industry and employee associations to assess gaps in policy and practice for the development of skills and to anticipate future needs. The project will address both technology-oriented high-end skills as well as middle and lower end skills to ensure that issues of inclusiveness are addressed in the context of rapidly growing economic sectors, including greening sectors through green jobs (Green jobs are jobs that reduce the environmental impact of enterprises and economic sectors, ultimately to levels that are sustainable as per ILO. Greening is a term for moving from traditional production processes or services to production processes or services that have a reduced environmental impact. For example —Manufacturing fuel-efficient cars, Solar Panels, Green-building construction work, Renewable energy products, Public Transport Operators, Smart electricity grids, Water and sanitation infrastructure, and green professionals such as green auditors.).

In order to assess the present scenario of skills development initiatives undertaken by TVET (Technical Vocational Education and Training) providers, we welcome your valuable inputs on how your training system is identifying skill gaps, developing courses to fill these gaps, placing trained students in jobs and following up with the employers/industry sectors to review programs. We would also like to understand the profile of students enrolling in your courses along with operational challenges you face to run such technical and vocational courses. We are enclosing a brief objective-type questionnaire which has seven sections, the last one being optional. It would be helpful if you could insert e-brochures/web links in the form or

R. Maclean et al., *Education and Skills for Inclusive Growth, Green Jobs and the Greening of Economies in Asia*, Technical and Vocational Education and Training: Issues, Concerns and Prospects 27, https://doi.org/10.1007/978-981-10-6559-0

email it to us separately. If you agree, we would also like to acknowledge your contribution by mentioning the name of your institution in the final report or you may choose to be anonymous. You may appropriately fill the last section at the end of the questionnaire. This information will be analyzed by the project researchers and will be treated as strictly confidential. We look forward to your response and hope that your inputs would help ADB and Govt. of India, develop robust and feasible policies and practices for the industry, training institutions, and markets which may complement each other for inclusive growth and green jobs.

Thank you,
Prof. Rupert Maclean
Chair Professor of International Education
The Education University of Hong Kong

If you have any questions about this survey or would like additional information please email: maclean@eduhk.hk

Q1. What is the Name of your Institution?

Q2. What is the Location/s of your Institution/s?

- ☐ Metropolitan
- ☐ Regional (Urban) or City
- ☐ Regional (Rural)
- ☐ All the above

Q3. How many students are currently enrolled in your institution? Example: 100

Q4. How many students are male? Example: 34

Q5. How many students are female? Example: 10

Q6. What is the age group of your students in years? Example: 18–25

Q7. What are the previous educational qualifications of students enrolling in your courses? (You can select more than one option, if applicable)

- ☐ Std. 5th Pass
- ☐ Std. 8th Pass
- ☐ High School

☐ Higher Secondary
☐ Graduate
☐ Postgraduate
☐ No Educational Qualifications
☐ Any Technical Certification
☐ Other:

Q8. How many total staff members are employed in your institution?

Q9. Out of the total number of staff members, how many are teachers/lecturers?

Q10. Out of the total number of staff members, how many are core workshop teachers/lecturers?

Q11. Out of the total number of staff members, how many are external/contract teachers/lecturers?

Q12. Out of the total number of staff members, how many are support staff (admin, mobilizers, accounts, etc.)?

Q13. Out of the total number of staff members, how many are females?

Q14. How do potential students find out about courses and programs offered by your institution? (You can choose more than one option)

☐ Family and friends (Word of Mouth)
☐ Former student(s)
☐ Career guidance in schools
☐ Career guidance outside schools
☐ Employer
☐ Road Shows and Public advertisement
☐ On line
☐ Other:

Q15. How important is work practice-based training in the courses/programs of your institution?

☐ Not Important
☐ Neutral
☐ Somewhat Important
☐ Very Important

Q16. How effective in your opinion are policies and practices of your institution in increasing the employability of its students and graduates? For example—Practices on Job Placement, On-the-job Training, etc.

☐ Not Effective
☐ Neutral
☐ Somewhat Effective
☐ Very Effective

Q17. How important does your institution rate student functional literacy/soft skills (like Spoken English, Basic IT, Life Skills) as part of its programs?

☐ Not Important
☐ Neutral
☐ Somewhat Important
☐ Very Important

Q18. Is your Institution a member of, or involved in, any of the following industry groups? (Please tick all that apply)

☐ Industry Association (e.g., CII, PHD Chamber, Retailer's Association of India, etc.)
☐ Professional Body (e.g., NSDC)
☐ Particular Employers
☐ None
☐ Other:

Q19. Can you name the Industry Associations and Professional Bodies you ticked in the previous question?

Q20. Has your institution been approached over the last year by any companies asking for training assistance for the following groups?

☐ Existing Workers
☐ New Workers
☐ Both of the above
☐ None of the above

Q21. Has your institution conducted surveys within the last year for any of the following groups in order to evaluate its courses and programs and in response to employer demands and labor force requirements?

- ☐ Enrolled Students
- ☐ Student Alumni
- ☐ Employers
- ☐ No Surveys Conducted
- ☐ Other:

Q22. What are the Industry Sectors your Skill Training Programs cater to?

- ☐ Construction/Building (cement, paints, masonry, carpentry, electrical, etc.)
- ☐ Energy and Power
- ☐ Transport/Automotive
- ☐ Hospitality
- ☐ Textile and Apparel
- ☐ Financial and Banking
- ☐ Consumer Durables (Plastic, Paper, Jute, Coir, etc.)
- ☐ Fast Moving Consumer Goods (Biscuits, Soap, Toiletries, etc.)
- ☐ Health Care (Patient Assistance)
- ☐ Metal and Mining (Pearl, Diamond, Aluminum, Graphite, etc.)
- ☐ Oil and Gas
- ☐ Retailing
- ☐ Telecom and Electronics
- ☐ Travel and Tourism
- ☐ Agriculture based (Dairy, Tobacco, Cotton, Sugar, Poultry, etc.)
- ☐ Other:

Q23. What are the types of certification provided to your students at the end of courses/training programs?

- ☐ Degree
- ☐ Masters
- ☐ Diploma
- ☐ PG Diploma
- ☐ Trade Skill Certificate
- ☐ Other:

Q24. Please provide information about each of your courses in the short format explained below—Course Name/Type of Certificate/Duration/Qualification of Students/MALES/FEMALES (Example—Basic Carpentry/Trade Certificate/6 Months/8th Pass to Higher Secondary/28 M/No F).

Q25. Does the institution regularly review and update its course and programs?

☐ No
☐ Not Sure, Don't Know
☐ Yes, on its own initiative
☐ Yes, as a response to external requests and directions

Q26. If you answered Yes to the previous question, how often are your courses reviewed?

☐ After each course/program delivery
☐ Once per semester
☐ Once per year
☐ Once in 2–4 years
☐ Every 5 or more years
☐ Other:

Q27. What is the main purpose of reviewing courses and programs?

☐ Quality control
☐ Meeting employers' requirements
☐ Responding to student satisfaction surveys
☐ Addressing the gap between graduate skills and labor market requirements
☐ Meeting institutional development needs
☐ Responding to government policies/guidelines
☐ Other:

Q28. In planning ahead to predict the labor market needs for workforce skills, does your institution use any of the following data sources or forecasting tools/mechanisms to make decisions about required training (types and content of course and programs?

☐ Institutional research (including feedback surveys)
☐ Published statistical data
☐ Government directions/advice
☐ Labor market analysis
☐ Skills observatories/skills councils
☐ Job announcements analysis
☐ Links with other TVET institutions
☐ Other:

Q29. Please state on a scale of 1–5, how you rate the strength of your institution's response to government initiatives with regard to economic development, environmental policies, skills development and green jobs. 1—Not Considered at all and 5—Very Closely Considered.

	1	2	3	4	5
Economic development					
Environmental policies					
Skills development					
Green jobs					

Q30. Please state on a scale of 1–5, the priority your institution gives to upskilling teachers and trainers to meet the needs of a greening economy? 1—Not Prioritized at all and 5—Highly Prioritized.

	1	2	3	4	5
Green upskilling of teachers					

Q31. Please let us know how strongly you agree or disagree with the following statements related to Institution's Approach to Economic Restructuring. Strongly Disagree (SD), Disagree (D), Unsure (U), Agree (A), Strongly Agree (SA).

	SD	D	U	A	SA
Over the last 3 years, economic development has influenced development of our programs					
Our institution is involved in initiatives with industry that influence development of training programs for required skills					
We have offered new courses/programs and closed others over the last 2 years					
There is increasing demand for a number of our courses/ programs					
Our current training programs are meeting skills demands at local or regional levels					
Recent program changes like course revisions and new courses reflect shifts from low skills (labor intensive) to middle-level skills jobs					
Recent program changes reflect shifts from low and mid-level skill jobs to high skills knowledge-based jobs.					
Our programs are guided by training/education standards of specific occupations					
Our programs are guided by skills profiles/occupational standards used in the industry					
Policies and practices in my institution are very effective for keeping graduates' skills and knowledge up to date					

Q32. Please let us know how strongly you agree or disagree with the following statements related to institution's approach toward green skills. Strongly Disagree (SD), Disagree (D), Unsure (U), Agree (A), Strongly Agree (SA).

	SD	D	U	A	SA
Institution needs to develop a full approach to adjust training to needs of greening economy					
Institution faces many challenges in addressing the development of skills needed for green jobs					
Institution is involved in internal and/or external partnership on green skills inclusion in courses/programs					
There is increasing demand for a number of our courses/ programs					
Institution needs to start new or continuing training programs to meet the demand for green skills					
New courses/programs have been developed to train students in "green" skills					
Additional skills and competencies have been added into existing courses to meet green skills requirements					
In this institution, "green" content is mainly learnt in "general" subjects					
In this institution, "green" content is mainly learnt in specialized technical subjects					
Green content is mainly learnt through industry exposure visits during the courses of the institution					
Comprehensive "green" skills standards have been developed in my institution					
A comprehensive certification system for "green" standards has been developed in my institution					
My institution is training students for industry/ies that are key drivers for economic growth in my region					
Scope of "green" job creation in my region is limited so it does not generate enough demand for green skills training					
My institution's policies and practices are very effective in developing the "green" skills and knowledge of our students so they are up to date to meet industry demands					

Q33. Changes in your programs/courses are the result of the following aspects of green economic changes: Strongly Disagree (SD), Disagree (D), Unsure (U), Agree (A), Strongly Agree (SA).

	SD	D	U	A	SA
Changes about greater use of energy from renewable sources (e.g., solar, wind, biomass, ocean, landfill gas, municipal solid waste, etc.)					
Changes about greater use of products and services that increase energy efficiency (energy-efficient equipment, appliances, vehicles, building design, etc.)					
Changes about greater use of processes that reduce/remove pollution and greenhouse gas emission and transition to recycling and reuse					
Changes about greater use of products and services that save natural resources (e.g., organic agriculture, sustainable forestry, land and soil management, water management)					
Changes about greater use of products and services that comply with environmental regulations and standards					

Q34. Please indicate below and rank the effectiveness of these teacher training methods at your institute.

	Not used	Not effective	Somewhat effective	Effective	Very effective
Individual mentoring by more experienced teacher					
Small group meetings/ discussions at the institute					
In-house seminars and courses by teachers themselves					
External trainer at the institute					
Online courses					
Industry organized training courses for your teachers at the institute					
Industry organized training courses for your teachers at the industry site					
Professional associations, skills councils					
Training arranged by teachers themselves					
Others					

Q35. Are teachers trained toward building green skills at your institution?

☐ Yes
☐ Not Yet

Q36. If you answered Yes to the previous question, please indicate below and rank the effectiveness of these teacher training methods toward green skills at your institution.

	Not used	Not effective	Somewhat effective	Effective	Very effective
Individual mentoring by more experienced teacher					
Small group meetings/ discussions at the institute					
In-house seminars and courses by teachers themselves					
External trainer at the institute					
Online courses					
Industry organized training courses for your teachers at the institute					
Industry organized training courses for your teachers at the industry site					
Professional associations, skills councils					
Training arranged by teachers themselves					
Others					

Q37. Do you have any further general comments to make?

Q38. What do you see as the outstanding opportunities offered by the development of education for inclusive economic growth and green jobs? (Can you provide any examples?)

Q39. What do you see as the outstanding challenges faced in the development of education for inclusive economic growth and green jobs? (Can you give any examples?)

Q40. Do you think that there are any particular issues faced by rural and regional institutions in providing skills development training, access, and opportunities?

Q41. What is your current position in the institution? Example—CEO, Program Manager, etc.

Q42. What is your age (years)? Example - 25

Q43. Are you Male or Female?

☐ Male
☐ Female

Q44. What are your qualifications? You can select multiple options below.

☐ Certificate
☐ Diploma
☐ Advanced Diploma
☐ Bachelor Degree
☐ Postgraduate Degree
☐ Other:

Q45. Would you like your organization's name to be acknowledged in the published report?

☐ Yes
☐ No

Q46. We invite you to participate in follow-up activities on some of the issues raised in this questionnaire. If you consent to be contacted as a potential participant (on an entirely voluntary basis), please give your contact details below. If you do not wish to participate, please ignore this invitation and no further contact will be made. Name/Phone Number/Email ID.

Appendix E
Employers' Interview Protocol

E.1 Interview Protocol with Industry Bodies

A key aspect of strong, sustainable, and inclusive growth is assuring adequate quality employment, which requires an adequate skills base. Durable links are needed between the world of work and training providers to match training provision with employment needs.

Sustaining economic growth without compromising the environment is a policy challenge that Asia has begun to address:

- the Asia-Pacific region uses three times the resources as the rest of the world to create one unit of GDP, and between 2000 and 2005 the resource intensity of its economy increased, in contrast to trends in the rest of the world.[1]
- the Asia-Pacific region, while emitting relatively low levels of greenhouse gases on a per-capita basis, is one of the fastest-growing sources of climate impacting emissions.[2]
- a number of studies estimate that by 2030, 100 million green jobs will be available worldwide—about 2% of the global workforce. Of these, 50 million green jobs are anticipated in Asia.

Enterprise surveys by the World Bank since 2000 reveal that 50% of firms surveyed in East Asia, Southeast Asia, and the Pacific are concerned with inadequate worker education and skills. Industry and business have initiated steps to train workers for green processes and products; however, governments, industries, and educational

[1]UN (2012a, b). *Low Carbon Green Growth Roadmap for Asia and the Pacific: Turning Resource Constraints and the Climate Crisis into Economic Growth Opportunities*. United Nations Economic and Social Commission for Asia and the Pacific. Korea International Cooperation Agency. Retrieved from: www.unescap.org/esd/environment/lcgg/.
[2]*Ibid.*

R. Maclean et al., *Education and Skills for Inclusive Growth, Green Jobs and the Greening of Economies in Asia*, Technical and Vocational Education and Training: Issues, Concerns and Prospects 27, https://doi.org/10.1007/978-981-10-6559-0

institutions have not yet developed coordinated actions to prepare the workforce in middle and higher green skills.

To support a coordinated approach to green growth in the region, the Asian Development Bank is funding this project, Education for Inclusive Growth and Green Jobs, in four countries, India, Sri Lanka, Viet Nam, and Indonesia. The intended impact of the research concerns improved and increased availability of job-oriented courses and skills training, including for green occupations.

Green jobs are jobs that reduce the environmental impact of enterprises and economic sectors, ultimately to levels that are sustainable.[3]

Greening is a term for moving from traditional production processes or services to production processes or services that have a reduced environmental impact.

(Have on cards which you let them have throughout the interview in case they need to refer back to the definitions?)

The information from this interview will be used for the sole purposes of the research project. Your comments will not be made public, if there may be a reason to link your comments to you, I will contact you first to seek your approval.

Do you mind if I record this interview solely for my purpose of developing comprehensive notes. No one else will listen to the recording.

Thank you for participating. This interview will take up to 2 hours.

E.2 Organizational Profile

1. What industry sectors does your organization cater to? Can you expand on the membership. Partners/clients breakdown per industry sector?
2. What services do you provide to your membership/partners/clients? (Training, lobbying, policy development, investment promotion, etc)
3. What kind of services around climate change/sustainability/carbon footprint reduction/greening business does your organization provide to your membership/partners/clients?
4. Does your organization work closely with other organizations that are involved in climate change/sustainability/carbon footprint reduction/greening business? If so, who and how do the organizations work together?
5. Is there a section or person dedicated to climate change/sustainability/carbon footprint reduction/greening business? If so what is their main focus over the next 12 months to 2 years?
6. Does the organization facilitate member networks, for example, an HR or training working group or a greening transitions working group? If so, what are these networks?

[3]ILO Greens Job definition.

7. Is the organization looking to change its services or membership based on climate change/sustainability/carbon footprint reduction/greening business issues?
8. Does the organization provide services to the value chain partners of the membership/clients that fall outside the industry sector?
9. Does your organization provide services to government? If so what is the nature of services? Do any of them relate to climate change/sustainability/greening business?

E.3 Forecasting—Helps Industry Develop a Coordinated Approach to Different Issues. The Following Section Prompts Discussion Around How we can Best Forecast Trends to Identify and Meet the Skills Development Needs of Industry more Actively.

10. What are some of the big issues that are facing the industry generally over the next 1 to 2 years?
11. What are your members identifying as the changing, or soon to change, business operational issues related to climate change/sustainability/carbon footprint reduction/greening business?
12. What are your members identifying as the changing, or soon to change, workforce/staff/contractor development needs to meet sustainability and greening challenges?
13. What approaches does the organization use to forecast trends and issues relevant to the membership/partners/clients?
14. Does the organization work with other organizations to forecast trends affecting the industry? If so, how?
15. Does the organization work with other organizations to identify and forecast the skills development needs of the industry? If so, how?
16. Has the organization been invited to participate in forecasting the skills development needs of industry? If so, by who and what did the participation involve?
17. Thinking about forecasting trends, are there any organizations, nationally or internationally, that standout as doing this well?
18. Thinking about forecasting skills development needs, are there any organizations, nationally or internationally, that standout as doing this well?
19. Has the organization discussed how forecasting could be improved? If so, what options have been discussed?
20. Has the organization discussed how forecasting skills development needs could be improved? If so, what options have been discussed?

E.4 Policy development—can Help Facilitate or Hinder Industry Growth and Involvement in a Number of Spheres within the Economy and Society. Generally, the following Section Prompts Discussion Related to Supporting Industry Leadership in Forecasting Skill Needs Generally with a Particular Emphasis on Green Jobs.

21. Are there particular government policies that are supporting industry involvement in the green economy/sustainability/climate change? Can you identify how these policies are affecting the industry?
22. Are there particular government policies that are hindering industry involvement in the green economy/sustainability/climate change? Can you identify how these policies are affecting the industry?
23. What processes does the organization use to influence policy? Can you cite an example?
24. What kinds of strategic processes does the organization use to seek to influence policy development, for example, the beginning of an electoral cycle or annual government priority setting exercises, to review, refine and implement new policy goals?
25. Does the organization participate in government forums on TVET training? If so, what are these forums and what is their mandate?
26. Does the organization have formal relationships with training organizations? Explain expand on the nature of the relationship.
27. Does the organization facilitate dialogue between its industry members and training providers to improve the relevance of training provision?
28. What internal resources does the organization need to be effective in influencing policy?
29. Thinking about skills development, are there organizations that you can work with/partner to influence the quality of training outcomes?
30. Are there examples where members are working with training organizations to improve the relevance of training provision for industry?

E.5 Skill Needs—for the Industry Sector Members

31. What new jobs, if any, do you envisage the sector will need in the future to meet the challenges of sustainability and greening?
32. Thinking about future trends and challenges facing the industry, what new skill sets do you think existing workers in the industry sector will require?
33. Which jobs do your members and your organization believe will change the most and why?

- Operators
- Technicians
- Service personnel
- Supervisors
- Administrators
- Management

34. Is the sector finding it difficult to recruit workers with the skills needed to meet greening and climate change challenges?
35. Is there a particular set of jobs that the sector is finding hardest to recruit in relation to green jobs? Please list them in order of importance to the sector
36. Overall according to the membership, what general or basic skills do workers lack?
37. What is the main effect of a lack of skilled workers in the sector?

- Not able to expand trade/services
- Not able to expand labor force
- Not as efficient as the sector could be
- Sectoral costs are higher than they should be
- The sector cannot meet sustainability targets/goals

38. Please tell me whether your organization/association provides training in green skills/green technologies to your members? This may be informal or formal
39. Does it involve the following:

- Member networks on greening and CC issue
- Lean Management, 3Rs
- Technology training
- Paid training programs through the organization/association
- Conferences or trade shows
- Industry association updates
- Speakers and experts
- Facilitation of government/industry meetings, roundtables
- Meritorious recognition for achievement in sustainability/green jobs

40. Are you aware of any current training courses conducted by enterprises in sustainability, the 3Rs, etc.?
41. Are you aware of any current training courses that could meet your industry sectors sustainability needs?
42. Do you think the sector could be responsive to flexible apprenticeships or internship for new or existing workers tailored specifically to your industry sector's needs?
43. What could be done by Government, other agencies, training providers to support your greening transition?
44. What coordination mechanisms could support the sector to successfully make the transition to a green economy?

Thank you for generously spending the time to share this information with us.

Appendix F
TVET Providers' Interview Protocol

F.1 Introduction

Thank you for agreeing to be interviewed. All information shared with me will be treated in the strictest confidence. Our report (based on the materials collected from different TVET providers) will not identify individuals or institutions: the published information will be anonymous.

This interview is a part of an Asian Development Bank project on Education and Skills for Growth and Green Jobs. *The aim of this interview is* to identify successful practices, opportunities, challenges, and requirements for training providers to develop training programs in addressing industry needs in skills development, with a specific focus on meeting the demands of climate change and environmental sustainability issues that industry is addressing.

Do you have any questions you would like to ask about the project?

In the interest of improved accuracy, would you mind if I tape record the interview and later transcribe some parts of it? I assure you that comments made will be treated in the strictest confidence.

F.2 Approaches Used by your Institution to Respond to Skills Needs

Please briefly describe your institution.

1. What is your role in the institution? Your responsibilities?
2. What are the main industries you are training for and what are the levels (low skills–high skills)?
3. Do you have links with industry? What sort of links? (e.g., workplace for training/industrial placement as a part of TVET courses; industry involved in accreditation panels; engaged in content development for courses; Allow site visits for students as part of their course; provide opportunities to gain industry experience for TVET teachers; engaged in skills policy formulation)

R. Maclean et al., *Education and Skills for Inclusive Growth, Green Jobs and the Greening of Economies in Asia*, Technical and Vocational Education and Training: Issues, Concerns and Prospects 27, https://doi.org/10.1007/978-981-10-6559-0

4. Does your institution receive any feedback from industry/employers about your graduates?
5. What other types of data (sources of information) is your institution using to make decisions about required training (types of courses and programs)?(e.g., types of statistical data, governments directions, and our research)
6. Does your institution use skills forecasting tools/mechanisms to predict the needs of the labor market (to anticipate skills needs) so you can update your courses/programs? (e.g., use of labor market information, involved with skills observatories or skills councils; job announcements' analysis) On a scale of 1–10, what is the priority given by your institution to skills forecasting and the matching of graduate skills to changing labor market needs?
7. Are there any procedures in place to review your courses/programs in response to industry/employer feedback received and information gained from forecasting tools? How often do you update your courses and programs?
8. What are the main opportunities and challenges in carrying out the above activities?

F.3 Approaches Used by your Institution to Respond to Government Policies/Initiatives That are Relevant to Skills Development

9. Does your institution try to respond to government initiatives that set up directions for *country's (region's) in terms of*
   - economic development;
   - environmental policies;
   - skills development policies;
   - if it does, what are the ways in which it responds?
10. Please give examples of recent policy documents in the areas of economic and social development, environment, and skills development that have influenced your institution's decisions about development of new courses/programs and updating existing ones?
11. What mechanisms, if any, has your institution used to translate policies into institutional practices)?
12. What are the main opportunities and challenges in trying to respond to government policy initiatives?

F.4 Approaches you Use to Respond to Economic Restructuring

13. What programs/courses are increasingly in demand in your institution? What are the reasons?
14. What programs/courses are declining in demand? What are the reasons?
15. To what extent do you think your institution meets changing skills demands at the local or regional levels? Could more be done? In what ways?
16. Do any recent changes in your programs reflect shifts from training for low skills jobs (labor intensive) to middle-level skills jobs? From low- and middle-level skills jobs to high skills jobs (knowledge-based, create opportunities for high-quality or high-wage jobs)?
17. In order to meet new skill demands what are your institution's plans in terms of courses/programs development? Revisions? Closures? Other activities?
18. What are the opportunities and challenges that relate to policies and practices in your institution for keeping skills and knowledge of your graduates up to date to meet industry demands?

F.5 Approaches your Institution Uses to Include Green Skills (Skills that are Required to Reduce the Environmental Impact of Enterprises and Economic Sectors, Ultimately to Levels that are Sustainable) in its Programs/Courses

19. What is the scope for green job creation in your region? Is it sufficient to generate needs for green skills training in institutions like yours?
20. Does your institution have, or is planning for, development of a comprehensive approach toward adjusting your training to the needs of a greening economy?
21. What are the main ways students are learning the "green content":
    - "general" subjects
    - "specialised" subjects
    - industry placement during the courses
    - other?
22. Has your institution implemented or planned changes in programs/courses as the result of the different aspects of green economic transitions: (*refer to the handout*)
23. If changes have been implemented or planned, do these relate to updating new skills and competences of students in existing courses or setting up new training programs/courses to meet the demand for green skills? Please give examples and the level of skills represented (Low, Middle or High).

24. What are the main reasons for adopting a changed approach?
    - changing natural or built environments;
    - policy and regulation;
    - technology and innovation; and
    - markets for green industries and consumer habits.
25. Does your institution use any skills forecasting tools to predict the needs of greening economies? On a scale of 1 to 10, what is the priority given by your institution to skills forecasting of "green jobs" and the matching of graduate skills to the needs of greening economies?
26. What are the roles of social partners (e.g., industry, government bodies, and associations) in adopting new approaches to training for "green jobs"?
27. Does your institution have/apply green skills standards? certification system for training courses and programs?
28. What are the opportunities and challenges faced by your institution in addressing the development of skills required for green jobs?

F.6 Approaches to Staff Training (Professional Development of Teachers and Workshop Teachers)

29. What guides teaching and learning in your institution's programs? (e.g., training/education standards for specific occupations; skills profiles/occupational standards used in the industry)
30. How is training provided in your institution to update skills and knowledge of general and workshop teachers? (refer to the handout)
31. What processes, if any, are in place or planned to support staff in their inclusion of "green" components in the training they provide?
32. Which of the ways of training you have outlined seems to be most effective, and why?
33. What are the main opportunities and challenges in helping teachers at your institution to update their skills and knowledge?

Are there any other comments you would like to make which are relevant to this study? If so, please feel free to make them. If you wish them to be off the record, I will stop the tape.

Thank you very much for your time and interest in being interviewed.

Bibliography and Further Reading List

Asian Development Bank. 2006. *Energy Efficiency and Climate Change Considerations for Road Transport in Asia*. Manila: Asian Development Bank. www.adb.org/Documents/CPSs/INO/2009/COBP-INO-2010-2012.pdf.

Asian Development Bank. 2009. *Focus on Education. Good Practice in Technical and Vocational Education and Training*. Manila: Asian Development Bank.

Asian Development Bank. 2010. *Education by 2020, A Sector Operations Plan*. Manila: Asian Development Bank.

Asian Development Bank. 2011. *Asia 2050, Realizing the Asian Century*. Manila: Asian Development Bank.

Asian Development Bank and Asian Development Bank Institute. 2012. *Policies and Practices for Low-Carbon Green Growth in Asia: Highlights*. Manila: ADB and Tokyo: ADBI. www.adb.org/publications/policies-and-practices-low-carbon-green-growthasia-highlights.

Association of Chartered Certified Accountants (ACCA). 2016. *Smarter cities, simpler cities Accounting for the city of the future*. http://www.accaglobal.com/content/dam/ACCA_Global/Technical/Future/pi-smarter-cities-simpler-cities.pdf.

Automotive Component Manufactures' Association. 2011. Auto News. Vol. 7, No.30. *Automotive Component Manufactures' Association News*. www.acma.in/acmanews/April-May2011/April-May-2011-issue.pdf.

Billett, S. 2004. Learning through Work: Workplace Participatory Practices. In H. Rainbird, A Fuller, and A Munro, eds. *Workplace Learning in C*ontext. London: Routledge.

Bird, J. and Lawton. 2009. *The Future's Green: Jobs and the UK Low-Carbon Transition, Institute of Public Policy Research*. www.ippr.org.

Capital Alliance. Various dates. *Sri Lanka Tourism Sector Summaries*. Colombo: Capital Alliance.

ChartsBin Statistics Collector Team. 2016. *Country Income Groups*. World Bank Classification. www.chartsbin.com/view/2438.

Commission of the European Communities. 2005. *Commission Staff Working Document on the Links between Employment Policies and Environment Policies*. 17 November. Brussels: SEC.

Confederation of Indian Industry. 2010. *Human Resource Development Strategies for the Indian Renewable Energy Sector. Final Report*. Delhi: Ministry of New and Renewable Energy, Government of India.

Eccles, R.G., Ioannou, I., and Serafeim, G. 2012. The Impact of a Corporate Culture of Sustainability on Corporate Behavior and Performance. *Harvard Business School Working Paper* No. 12-035. www.trippel.sdg.no/wp-content/uploads/2014/09/Eccles-HBR_The-Impact-of-a-Corporate-Culture-of-Sustainability1.pdf Economic and Social Statistics of Sri Lanka. 2013. Colombo: Government of Sri Lanka. www.statistics.gov.lk/.

Educational Consultants India Limited. 2005. Studying the Effectiveness of Vocational Training in the Private Sector in India: Analysis of Data in Eight States. *Working Paper Prepared for the World Bank.*

R. Maclean et al., *Education and Skills for Inclusive Growth, Green Jobs and the Greening of Economies in Asia*, Technical and Vocational Education and Training: Issues, Concerns and Prospects 27, https://doi.org/10.1007/978-981-10-6559-0

EMCO Reports. 2010. *Towards A Greener Labour Market—The Employment Dimension of Tackling Environmental Challenges*, s.l.: s.n. programmaleonardo.net/llp/green_job/documenti/EMCO_Green%20economy.pdf.

Environmental Business International (EBI). 2012. Global Environmental Markets. *Environmental Business Journal*. (25): 6 and 7.

Environmental Health News. 2012. *Environmental Health Sciences*. www.environmentalhealthnews.org/ehs/news/2012/unequal-exposures.

Esposto, Alexis S. 2016. *Skills needs in emerging green jobs in the building and tourism industries in Thailand*. Bangkok: ILO.

European Centre for the Development of Vocational Training. 2009. *Future Skill Needs for the Green Economy*. Luxembourg: Publications Office of the European Union. www.cedefop.europa.eu/EN/publications/5025.aspx.

European Centre for the Development of Vocational Training. (2010a). *Skills for Green Jobs—European Synthesis Report*, Luxembourg: Publications Office of the European Union.

European Centre for the Development of Vocational Training. (2010b). Skills for Green Jobs. *Briefing Note*. Thessaloniki: European Centre for the Development of Vocational Training.

European Centre for the Development of Vocational Training. 2012.Green Skills: An Environmental Awareness in Vocational Education and Training. *Research paper* #24. Luxembourg: Publications Office of the European Union.

European Commission. 2013. *European Employment Observatory Review—Promoting Green Jobs throughout the Crisis: A Handbook of Best Practices in Europe*, Directorate-General for Employment, Social Affairs and Inclusion. Luxembourg: Publications Office of the European Union. www.eu-employment-observatory.net/explore.aspx?document=1358.

Ewing B., Moore, D., Goldfinger, S., Oursler, A.,Reed, A, and Wackernagel, M. 2010. *The Ecological Footprint Atlas 2010*. Oakland: Global Footprint Network.

Frankfurt School-UNEP Centre/BNEF. 2016. *Global Trends in Renewable Energy Investment 2016*. http://www.fs-unep-centre.org.

Global Footprint Network. 2012. Ecological Footprint per Capita. *Ecological Wealth of Nations*. www.footprintnetwork.org/ecological_footprint_nations/ecological_per_capita.html.

Global Footprint Network. 2014. *Living Planet Report, 2014: Facts*. www.footprintnetwork.org/en/index.php/gfn/page/living_planet_report_2014_facts.

Goetsch, D.L., Davis, S.B. (Eds.). 2009. *Quality Management for Organizational Excellence: Introduction to Total Quality*. USA: Pearson.

Government of India, Planning Commission. 2011. *Approach Paper to 12th Five Year Plan—Faster, Sustainable and More Inclusive Growth*. Delhi. http://planningcommission.gov.in/plans/planrel/12appdrft/appraoch_12plan.pdf.

Government of Indonesia. 2013. Aligning the New Curriculum. *Government Regulation No. 70*. Jakarta.

Government of Viet Nam. 2005. *Law on Environmental Protection*. Ha Noi: Viet Nam Government Portal. Article 107. http://vea.gov.vn/en/laws/LegalDocument/Pages/LawNo52_2005_QH11onenvironmentalprotection.aspx.

Government of Viet Nam. 2008. *National Strategy on Climate Change*. Ha Noi: Viet Nam Government Portal.

Government of Viet Nam. 2012a. *National Green Growth Strategy for the Period 2011–2020 with a Vision to 2050*. Ha Noi: Viet Nam Government Portal.

Government of Viet Nam. 2012b. *National Strategy on Environmental Protection to 2020 with Vision to 2030*. Ha Noi: Viet Nam Government Portal.

Government of Viet Nam. 2012c. *Viet Nam Sustainable Development Strategy for 2011–2020*. Ha Noi: Viet Nam Government Portal.

Government of Viet Nam. (Draft). Green Industry Policy Framework. Ha Noi: Viet Nam Government Portal.

Green Climate Fund (GCF). 2017. *Green Climate Fund*. http://www.greenclimate.fund/home.

Green Climate Fund (GCF) Readiness Programme. 2017. *GCF Readiness Programme*. http://www.gcfreadinessprogramme.org.

Green Skills Agreement: An Agreement between the Australian Government and the State and Territory Governments http://www.ivet.com.au/cgi-bin/user.pl?download_file=1&file=17.

Hager, P. 1997. *Learning in the Workplace*: Review of Research. National Centre for Vocational Educational Research. Leabrook, Australia.

Hellström, Tomas. Programme on Innovation, Higher Education and Research for Development (IHERD) Background Document, Draft Synthesis Report. Sweden: Lund University. https://www.oecd.org/sti/Draft_OECD%20synthesis%20report_final.pdf.

Her Majesty's Government, United Kingdom. 2011. Skills for a Green Economy, *A Report on the Evidence, A Joint Her Majesty Government Report with Contributions from the Department for Business, Innovation and Skills, the Department of Energy and Climate Change, and the Department for Environment, Food and Rural Affairs*. www.gov.uk/government/uploads/system/uploads/attachment_data/file/32373/11-1315-skills-for-a-green-economy.pdf.

ICRA Lanka. 2011. *Construction Industry Report 2011. ICRA Management Consulting Services Limited*. www.icralanka.com/Sri%20Lanka%20Construction%20-%20Sept%2015%20final.pdf.

Indian Railways, Government of India. 2012. *Indian Railways—Fact and Figure 2011–12*. www.indianrailways.gov.in/railwayboard/uploads/directorate/stat_econ/pdf/FACTS_FIGURES_ENG_2011_12/13.pdf.

International Council for Science (ICSU) and International Social Science Council (ISSC). 2015. Review of the Sustainable Development Goals: The Science Perspective. Paris: ICSU.

Indonesia Ecotourism Network. 2014. *Background*. www.indecon.or.id/?q=node/42.

International Labour Organization. 2008. *Skills for Improved Productivity, Employment Growth and Development*. Geneva, Switzerland: Report for the International Labour Conference, 97th Session.

International Labour Organization. 2011a. *Skills for Green Jobs—A Global Review. Synthesis Report Based on 21 Country Studies*. Geneva, Switzerland: ILO Skills and Employability Department. http://www.ilo.org/wcmsp5/groups/public/@dgreports/@dcomm/@publ/documents/publication/wcms_159585.pdf.

International Labour Organization. 2011b. *Towards a Greener Economy: The Social Dimensions*. Geneva: International Institute for Labour Studies.

International Labour Organization. 2011c. Skills and Occupational Needs in Renewable Energy. Geneva.

International Labour Organization. 2012a. *Employment Creation Model in Indonesia*: Promoting Green Jobs and Livelihoods in Sustainable Tourism. Ecotourism and Green Homestays.

International Labour Organization. 2012b. *Working Towards Sustainable Development – Opportunities for Decent Work and Social Inclusion in a Green Economy*. Geneva: International Labour Organization.

International Labour Organization. 2013. *Green jobs becoming a reality: Progress and outlook*. 2013. Geneva.

International Labour Organization. 2015a. *Guidelines for a just transition towards environmentally sustainable economies and societies for all*. Geneva. http://www.ilo.org/wcmsp5/groups/public/—ed_emp/–emp_ent/documents/publication/wcms_432859.pdf.

International Labour Organization. 2015b. Undated. Skills and Employability. *Employment for Social Justice and a Fair Globalization Overview of ILO Programmes*. ILO http://www.ilo.org/wcmsp5/groups/public/@ed_emp/documents/publication/wcms_140945.pdf.

International Labour Organization/International Organization of Employers/United Nations Environment Programme/International Trade Union Conference. 2008. *Green Jobs: Towards Decent Work in a Sustainable, Low-Carbon World*. United Nations Environment Programme. www.unep.org/labour_environment/features/greenjobs.asp.

International Labour Organization and European Union. 2011a. *Comparative Analysis of Methods of Identification of Skill Needs on the Labour Market in Transition to the Low Carbon Economy*. Geneva: International Labour Organization.

International Labour Organization and European Union. 2011b. *Investment in renewable energy generates jobs. Supply of skilled workforce needs to catch up*. Geneva: International Labour Office and European Union.

International Renewable Energy Agency (IRENA). 2014. *Renewable Energy and Jobs: Annual Review 2014*. Abu Dhabi: IRENA. http://www.irena.org/Publications/rejobs-annual-review-2014.pdf.

International Labour Organization and European Union. 2015. *Renewable Energy Benefits: Measuring the Economics*. Abu Dhabi: IRENA.

International Labour Organization and European Union. 2016. Renewable Energy Benefits: Measuring the Economics. Abu Dhabi: IRENA. http://www.irena.org/DocumentDownloads/Publications/IRENA_Measuring-the-Economics_2016.pdf.

International Trade Centre. 2014. *Trade in Environmental Goods and Services: Opportunities and Challenges*. Geneva: Intracen. http://www.intracen.org/publication/Trade-in-environmental-goods-and-services-Opportunities-and-challenges/.

Keivani, R., Tah, Joseph H.M., Kurul, E. and Abanda, H. 2010. *Green Jobs Creation Through Sustainable Refurbishment in the Developing Countries*. Geneva: International Labour Organization.

Klasen, S. 2010. Measuring and Monitoring Inclusive Growth: Multiple Definitions, Open Questions, and Some Constructive Proposals. *ADB Sustainable Development Working Paper Series*. Manila: Asian Development Bank.

Maclean, R., Jagannathan, S., and Sarvi, J. (Eds.). 2012. *Skills Development for Inclusive and Sustainable Growth in Developing Asia-Pacific*. Dordrecht: Springer.

Maclean, R., Jagannathan, S., and Sarvi, J. (Eds.). 2014 Skills for Inclusive and Sustainable Development: Perspectives from the Asia-Pacific Region and Beyond Prospects*: Quarterly Review of Comparative Education. (Special Issue)*. (XLIV): 2, June. Geneva: International Bureau of Education.

Martinez-Fernandez, C. and Powell, M. 2009. *Employment Skills Strategies in Southeast Asia: Setting the Scene*. Paris: Organisation for Economic Co-operation and Development, Local Economic and Employment Development. https://www.oecd.org/employment/leed/45592893.pdf.

McKinsey and Company. 2017. *Moving toward a Circular Economy*. http://www.mckinsey.com/business-functions/sustainability-and-resource-productivity/our-insights/moving-toward-a-circular-economy.

Mourshed, M., Farrell, D., and Barton, D. 2012. *Education to Employment: Designing a System that Works*. McKinsey Centre for Government. www.mckinseyonsociety.com/education-to-employment.

NCWE and AED. 2011. *Going Green: The Vital Role of Community Colleges in Building a Sustainable Future and Green Workforce*. www.ncwe.org.

National Skills Development Corporation. 2009. *Human Resource and Skill Requirements in the Transportation, Logistics, Warehousing and Packaging Sector*. New Delhi, India: National Skills Development Corporation.

O'Dell, Kathleen, Peters, Sophia, and Wharton, Kate. Women, Energy, and Economic Empowerment. Boiling Point: *A Practitioner's Journal on Household Energy, Stoves and Poverty Reduction*. (66): 2015. HEDON Household Energy Network. http://www.energia.org/cms/wp-content/uploads/2015/07/BP66-Women-Energy-and-Economic-Empowerment-.compressed.pdf.

Organisation for Economic Co-operation and Development. 2010a. *Cities and Climate Change*. Paris: Organisation for Economic Co-operation and Development Publishing.

Organisation for Economic Co-operation and Development. 2010b. *Measuring Innovation – A New Perspective*. Paris: Organisation for Economic Co-operation and Development Publishing.

Organisation for Economic Co-operation and Development. 2010c. *Greening Jobs and Skills: Labour Market Implications of Addressing Climate Change, OECD Local Economic and Employment Development (LEED) Working Document.* https://www.oecd.org/cfe/leed/44683169.pdf.

Organisation for Economic Co-operation and Development. 2011. *Towards green growth: A summary for policy makers*. Brochure prepared for the OECD Meeting of the Council at Ministerial Level. 25–26 May. Paris https://www.oecd.org/greengrowth/48012345.pdf.

Organisation for Economic Co-operation and Development. 2012. *The Jobs Potential of a Shift Towards a Low-carbon Economy*. Final Report for the European Commission. Paris: DG Employment. Division for Employment Analysis and Policy. www.oecd.org/els/emp/50503551.pdf.

Organisation for Economic Co-operation and Development. 2014. *Focus on Inequality and Growth*. Paris. www.oecd.org/els/soc/Focus-Inequality-and-Growth-2014.pdf.

Organisation for Economic Co-operation and Development. 2015. *In It Together: Why Less Inequality Benefits All*. Paris. www.oecd.org/els/soc/OECD2015-In-It-Together-Chapter1-Overview-Inequality.pdf.

Organisation for Economic Co-operation and Development. 2017. *Investing in Climate, Investing in Growth*. http://www.oecd.org/environment/investing-in-climate-investing-in-growth-9789264273528-en.htm.

Piaget, J. 1978. *Behaviour and Evolution*. Trans. D. N. Smith. New York: Pantheon Books.

Platform for Advancing Green Human Capital. 2016. *Green Jobs, Skills, Education and Training. Reflections from COP21: Inspiring Dialogues on the Road to COP22*. www.unevoc.unesco.org/up/COP21brochure.pdf.

Pro Enviro. 2008. *Skills for a Low-carbon and Resource Efficient Economy (LCREE). A Review of the Evidence*. Report for Department of Environment, Food and Rural Affairs. United Kingdom. www.archive.defra.gov.uk/environment/business/scp/documents/LCREE-final-report.pdf.

Ranieri, R. and Ramos, R.A. 2013. Inclusive Growth: The Building up of a Concept, *IPC-IG Working Paper* No. 104. Brasília, Brazil: International Policy Centre for Inclusive Growth. www.ipc-undp.org/pub/IPCWorkingPaper104.pdf.

Sambijantoro, Satria. 2015. Esemka national car project wins Jokowi's backing. *The Jakarta Post* 27 February 2015. www.thejakartapost.com/news/2015/02/27/esemka-national-car-project-wins-jokowi-s-backing.html.

Slingenberg, A., Rademaekers, K., Sincer, E., and van der Aa, R. 2008. Environment and Labour Force Skills. *Overview of the Links Between the Skills Profile of the Labour Force and Environmental Factors*. Brussels: (ECORYS) European Commission DG Environment.

Smith, B. 2012. *Integrating Work Experience in Technical Vocational Education and Training in Schools*. International Conference on Integration of Academic Courses with Vocational Education in Secondary Schools. Delhi: National Institute of Open Schooling.

Sri Lanka Sustainable Energy Authority. www.energy.gov.lk/sub_pgs/energy_managment_regulatory_1_2.html#4.

Stoevska, V. and Hunter, D. 2012. Proposals for the Statistical Definition and Measurement of Green Jobs. Preliminary Paper Prepared for Discussion by the London Group on Environmental-Economic Accounts. Geneva: International Labour Organization. www.unstats.un.org/unsd/default.htm.

Sustainable Development Knowledge Platform. Green growth. https://sustainabledevelopment.un.org/index.php?menu=1447.

Technopark—Confederation of Indian Industries. 2009. Case for Setting up Sector Skill Councils in India. In *CII National Conference on Education: Linking Education to Employability*. Delhi. 21 April 2009. http://rasci.in/downloads/2009/Linking_Education_Employability_2009.pdf.

Tertiary and Vocational Education Commission. 2012. *Corporate Plan 2013–2017*. Colombo. http://www.tvec.gov.lk/pdf/corporat_plan_2013_2017.pdf.

Trading Economics. 2014a. *GDP Annual Growth Rate—Countries – List*. New York. www.tradingeconomics.com/country-list/gdp-annual-growth-rate.

Trading Economics. 2014b. *Viet Nam GDP Growth Rate 2000–2014*. New York. http://www.tradingeconomics.com/vietnam/gdp-growth.

United Nations. 2012a. Economic and Social Survey of Asia and Pacific (ESCAP), *Pursuing Shared Prosperity in an Era of Turbulence and High Commodity Prices*. New York: United Nations Publication. www.unescap.org/pdd/publications/survey2012/download/Survey_2012.pdf.

United Nations. 2012b. Low Carbon Green Growth Roadmap for Asia and the Pacific: *Turning Resource Constraints and the Climate Crisis into Economic Growth Opportunities*. United Nations Economic and Social Commission for Asia and the Pacific. Seongnam: Korea International Cooperation Agency. www.unescap.org/esd/environment/lcgg/.

United Nations. 2015. *Transforming Our World: the 2030 Agenda for Sustainable Development*. New York. https://sustainabledevelopment.un.org/post2015/transformingourworld.

United Nations Capital Development Fund. 2013. Inclusive Future—Inequality, Inclusive Growth and the Post-2015 Framework. *UNCDF Policy Notes*. New York: United Nations Capital Development Fund.

United Nations Department of Economic and Social Affairs (UNDESA). 2012. *A guidebook to the Green Economy*. Division for Sustainable Development, UNDESA. https://sustainabledevelopment.un.org/content/documents/GE%20Guidebook.pdf.

United Nations Development Programme. 2011. *India Fact Sheet: Economic and Human Development Indicators*. New York: United Nations Development Programme. www.undp.org/content/dam/india/docs/india_factsheet_economic_n_hdi.pdf.

United Nations Development Programme, United Nations Department of Economic and Social Affairs and International Labour Organization. 2012. *Jobs, Decent Work and Inclusive Growth*. Report of the E-Discussion. New York: United Nations. www.un.org/en/ecosoc/newfunct/amredis2012.shtml.

United Nations Development Programme, United Nations Department of Economic and Social Affairs and International Labour Organization. 2010a. *The World's Women 2010: Trends and Statistics*. New York: United Nations.

United Nations Development Programme, United Nations Department of Economic and Social Affairs and International Labour Organization. 2010b. *Sustainable Development Goals*. UNDP website. http://www.undp.org/content/undp/en/home/sustainable-development-goals.html.

United Nations Economic Commission for Europe, International Labour Organization, International Organisation of Employers and International Trade Union Confederation. 2008. *Green Jobs: Towards Decent Work in a Sustainable, Low Carbon World*. Nairobi: United Nations Office at Nairobi (UNON). http://www.ilo.org/wcmsp5/groups/public/—dgreports/—dcomm/documents/publication/wcms_098504.pdf.

United Nations Economic and Social Commission for Asia and the Pacific. 2015. *Regional Trends Report on Energy for Sustainable Development in Asia and the Pacific 2015*. Bangkok: ESCAP.

United Nations Economic and Social Commission for Asia and the Pacific. 2012. *Green Growth and Green Economy*. http://www.unescap.org/our-work/environment-development/green-growth-green-economy/about.

United Nations Educational, Scientific and Cultural Organization. 2011. *Education for Sustainable Development, Country Guidelines for Changing the Climate of Teacher Education to Address Sustainability: Putting Transformative Education into Practice —Brunei Darussalam, Indonesia, Malaysia, Philippines and Timor-Leste*. Jakarta, Indonesia.

United Nations Educational, Scientific and Cultural Organization (UNESCO). 2016a. *Global Education Monitoring Report. Inclusive and Sustainable Cities*. Paris.

United Nations Educational, Scientific and Cultural Organization (UNESCO). 2016b. UNESCO Website. *Basic Sciences, Center of Excellence*. http://www.unesco.org/new/en/natural-sciences/science-technology/basic-sciences/international-basic-sciences-programme/center-of-excellence/.

United Nations Environment Programme—United Nations Economic Commission for Europe. 2008. *Green Jobs: Towards Decent Work in a Sustainable Low Carbon World*. Nairobi: United Nations Environment Programme.

United Nations Environment Programme. 2008. *Green Jobs: Towards Decent Work in a Sustainable, Low-carbon World*. Washington, DC: Worldwatch Institute.

United Nations Environment Programme. 2010a. *A Brief for Policymakers on the Green Economy and Millennium Development Goals*. Geneva.

United Nations Environment Programme. 2010b. *Green Economy Developing Countries Success Stories*. Geneva. http://www.unep.org/pdf/GreenEconomy_SuccessStories.pdf.

United Nations Environment Programme. 2011a. *Towards a Green Economy: Pathways to Sustainable Development and Poverty Eradication*. www.unep.org/greeneconomy.

United Nations Environment Programme. 2011b. *Green Economy Report*. http://web.unep.org/greeneconomy/sites/unep.org.greeneconomy/files/field/image/green_economyreport_final_dec2011.pdf.

United Nations Environment Programme. 2016a. *Global Environmental Outlook. Regional Assessment for Asia and the Pacific*. Nairobi: United Nations Environment Programme. file:///C:/Users/Rhona/Downloads/-GEO-6__Global_Environment_Outlook__Regional_assessment_for_Asia_and_the_Pacific-2016GEO-6_Global_Environment_Outlook_Regional_assessment_for_Asia_and.pdf.

United Nations Environment Programme. 2016b. *Global Gender and Environment Outlook*. Nairobi: United Nations Environment Programme. file:///C:/Users/Rhona/Downloads/-Global_gender_and_environment_outlook-2016gender_global_and__environment_outlook_2016.pdf.pdf.

United Nations Industrial Development Organization and Global Green Growth Institute. 2015. *Global Green Growth: Clean Energy Industrial Investments and Expanding Job Opportunities*. http://www.unido.org/fileadmin/user_media_upgrade/Resources/Policy_advice/GLOBAL_GREEN_GROWTH_REPORT_vol1_final.pdf.

United Nations System Task Team on the Post 2016 UN Development Agenda. 2015. *Science, innovation and technology for sustainable development in the global partnership beyond*. http://www.un.org/en/development/desa/policy/untaskteam_undf/thinkpieces/28_thinkpiece_science.pdf.

World Bank. 2010. *Securing the Present, Shaping the Future. East Asia and Pacific Economic Outlook*. Vol. 1. Washington, DC: The World Bank.

World Bank. 2012. Inclusive Green Growth. Washington, DC: World Bank. http://siteresources.worldbank.org/EXTSDNET/Resources/Inclusive_Green_Growth_May_2012.pdf.

World Wildlife Fund. 2014. *Living Planet Report 2014: Species and Spaces, People and Places*. Gland: World Wildlife Fund. www.wwf.or.jp/activities/lib/lpr/WWF_LPR_2014.pdf.

Index

R. Maclean et al., *Education and Skills for Inclusive Growth, Green Jobs and the Greening of Economies in Asia*, Technical and Vocational Education and Training: Issues, Concerns and Prospects 27, https://doi.org/10.1007/978-981-10-6559-0

The manufacturer's authorised representative in the EU is Springer
Nature Customer Service Centre GmbH, Europaplatz 3, 69115 Heidelberg,
Germany. If you have any concerns regarding our products, please
contact ProductSafety@springernature.com

Printed and bound by CPI Group (UK) Ltd, Croydon, CR0 4YY
30/09/2026
02225446-0001